Recent Advances in Surface and Coatings Technology

Recent Advances in Surface and Coatings Technology

Selected Articles Published by MDPI

MDPI • Basel • Beijing • Wuhan • Barcelona • Belgrade • Manchester • Tokyo • Cluj • Tianjin

This is a reprint of articles published online by the open access publisher MDPI (available at: http://www.mdpi.com). The responsibility for the book's title and preface lies with Antonia Nostro.

For citation purposes, cite each article independently as indicated on the article page online and as indicated below:

LastName, A.A.; LastName, B.B.; LastName, C.C. Article Title. *Journal Name* **Year**, *Article Number*, Page Range.

ISBN 978-3-03928-404-7 (Pbk)
ISBN 978-3-03928-405-4 (PDF)

Contents

Preface to "Recent Advances in Surface and Coatings Technology" vii

Lorenzo Lisuzzo, Giuseppe Cavallaro, Filippo Parisi, Stefana Milioto, Rawil Fakhrullin and
Giuseppe Lazzara
Core/Shell Gel Beads with Embedded Halloysite Nanotubes for Controlled Drug Release
Reprinted from: *Coatings* **2019**, *9*, 70, doi:10.3390/coatings9020070 . . . . **1**

Daniela Predoi, Simona Liliana Iconaru, Mihai Valentin Predoi, Nicolas Buton,
Christelle Megier and Mikael Motelica-Heino
Biocompatible Layers Obtained from Functionalized Iron Oxide Nanoparticles in Suspension
Reprinted from: *Coatings* **2019**, *9*, 773, doi:10.3390/coatings9120773 . . . . **9**

Domenico Sagnelli, Robert Cavanagh, Jinchuan Xu, Sadie M. E. Swainson,
Andreas Blennow, John Duncan, Vincenzo Taresco and Steve Howdle
Starch/Poly (Glycerol-Adipate) Nanocomposite Film as Novel Biocompatible Materials
Reprinted from: *Coatings* **2019**, *9*, 482, doi:10.3390/coatings9080482 . . . . **25**

Andreas Breitwieser, Philipp Siedlaczek, Helga Lichtenegger, Uwe B. Sleytr and
Dietmar Pum
S-Layer Protein Coated Carbon Nanotubes
Reprinted from: *Coatings* **2019**, *9*, 492, doi:10.3390/coatings9080492 . . . . **37**

Victor Gomes Lauriano Souza, João Ricardo Afonso Pires, Carolina Rodrigues,
Patricia Freitas Rodrigues, Andréia Lopes, Rui Jorge Silva, Jorge Caldeira, Maria Paula
Duarte, Francisco Braz Fernandes, Isabel Maria Coelhoso and Ana Luisa Fernando
Physical and Morphological Characterization of Chitosan/Montmorillonite Films Incorporated
with Ginger Essential Oil
Reprinted from: *Coatings* **2019**, *9*, 700, doi:10.3390/coatings9110700 . . . . **51**

Mayra Sapper, Lluís Palou, María B. Pérez-Gago and Amparo Chiralt
Antifungal Starch–Gellan Edible Coatings with Thyme Essential Oil for the Postharvest
Preservation of Apple and Persimmon
Reprinted from: *Coatings* **2019**, *9*, 333, doi:10.3390/coatings9050333 . . . . **71**

Beata Majkowska-Marzec, Dorota Rogala-Wielgus, Michał Bartmański, Bartosz Bartosewicz
and Andrzej Zieliński
Comparison of Properties of the Hybrid and Bilayer MWCNTs—Hydroxyapatite Coatings on
Ti Alloy
Reprinted from: *Coatings* **2019**, *9*, 643, doi:10.3390/coatings9100643 . . . . **87**

Clare Lubov Donaghy, Ryan McFadden, Graham C. Smith, Sophia Kelaini, Louise Carson,
Savko Malinov, Andriana Margariti and Chi-Wai Chan
Fibre Laser Treatment of Beta TNZT Titanium Alloys for Load-Bearing Implant Applications:
Effects of Surface Physical and Chemical Features on Mesenchymal Stem Cell Response
and *Staphylococcus aureus* Bacterial Attachment
Reprinted from: *Coatings* **2019**, *9*, 186, doi:10.3390/coatings9030186 . . . . **101**

Danaja Štular, Matic Šobak, Mohor Mihelčič, Ervin Šest, Ilija German Ilić, Ivan Jerman,
Barbara Simončič and Brigita Tomšič
Proactive Release of Antimicrobial Essential Oil from a "Smart" Cotton Fabric
Reprinted from: *Coatings* **2019**, *9*, 242, doi:10.3390/coatings9040242 . . . . **121**

Lavinia-Florina Călinoiu, Bianca Eugenia Ştefănescu, Ioana Delia Pop, Leon Muntean and
Dan Cristian Vodnar
Chitosan Coating Applications in Probiotic Microencapsulation
Reprinted from: *Coatings* **2019**, *9*, 194, doi:10.3390/coatings9030194 . . . . **139**

Preface to "Recent Advances in Surface and Coatings Technology"

Development of new surfaces and coatings has resulted in scientific and technological progress. Coatings technology offers the possibility of producing advanced surface with tailored physicochemical and biological properties to meet requirements for food, biomedical, and environmental applications. This e-book gathers recently published papers that have attracted considerable attention and consideration, providing updates on this research topic. Specifically, this book collects new information on designed surfaces, coatings, and materials based on metals, polymers, ceramics, and composites. Some research describes the techniques for the preparation and characterization of gel, functional surfaces, and coatings, and their potential application in drug delivery or films for food preservation. Other studies outline different surface preparation processes or novel techniques for modifying the surface features for load-bearing implant applications. One paper focuses on the synthesis of stimuli-responsive hydrogels for new perspectives such as antimicrobial coatings for cotton–cellulose fabric. A review is devoted to probiotic microencapsulation technology. The collection delivers scientific findings to both academic and industrial readers.

Antonia Nostro

Communication

Core/Shell Gel Beads with Embedded Halloysite Nanotubes for Controlled Drug Release

Lorenzo Lisuzzo [1], Giuseppe Cavallaro [1,2], Filippo Parisi [1], Stefana Milioto [1,2], Rawil Fakhrullin [3] and Giuseppe Lazzara [1,2,*]

[1] Dipartimento di Fisica e Chimica, Università degli Studi di Palermo, Viale delle Scienze, pad. 17, 90128 Palermo, Italy; lorenzo.lisuzzo@unipa.it (L.L.); giuseppe.cavallaro@unipa.it (G.C.); filippo.parisi@unipa.it (F.P.)

[2] Consorzio Interuniversitario Nazionale per la Scienza e Tecnologia dei Materiali, INSTM, Via G. Giusti 9, I-50121 Firenze, Italy; stefana.milioto@unipa.it

[3] Institute of Fundamental Medicine and Biology, Kazan Federal University, Kreml uramı 18, Kazan, Republic of Tatarstan 420008, Russia; kazanbio@gmail.com

* Correspondence: giuseppe.lazzara@unipa.it; Tel.: +39-091-2389-7962

Received: 15 November 2018; Accepted: 22 January 2019; Published: 24 January 2019

Abstract: The use of nanocomposites based on biopolymers and nanoparticles for controlled drug release is an attractive notion. We used halloysite nanotubes that were promising candidates for the loading and release of active molecules due to their hollow cavity. Gel beads based on chitosan with uniformly dispersed halloysite nanotubes were obtained by a dropping method. Alginate was used to generate a coating layer over the hybrid gel beads. This proposed procedure succeeded in controlling the morphology at the mesoscale and it had a relevant effect on the release profile of the model drug from the nanotube cavity.

Keywords: halloysite; alginate; chitosan; gel beads; drug release

---

1. Introduction

Researchers have defined hydrogels in many different ways, but nowadays the most accepted definition is the existence of a three-dimensional network, formed by the cross-linking of polymeric chains, that possesses the capability to swell thanks to the presence of hydrophilic groups and to maintain a very high amount of water in its structure [1,2]. Since their discovery, hydrogels have received attention from the scientific community due to the wide range of applications they can be used for: Environmental issues like water remediation, drug delivery systems and tissue engineering, cosmetic and food packaging industry, and oil spill recovery [3–8]. Furthermore, with the evolution of nanotechnology, the challenge to design and prepare hydrogels with specific and requested features at the nano-scale led to the development of nanohydrogels. Among the different polymeric species that can be used to achieve this aim, polysaccharides cover a marked importance, especially in the preparation of the so-called "polysaccharide-based natural hydrogels", for some of their most peculiar properties such as water solubility and swelling capacity, biocompatibility and biodegradability, self-healing and pH sensitivity that are crucial for their use [9,10]. Moreover, the possibility to modify the structure of the polysaccharides and the adaptability of their networks allows for the development of eco-friendly smart materials [11,12]. To date, the most widely used raw materials include natural biopolymers such as chitosan, alginate, pectin and cellulose. One of the major factors limiting the use of nanohydrogels is their structural instability, thus making necessary the use, among others, of inorganic nanoparticles to overcome them [13–16].

Among clays, halloysite nanotubes (HNTs) have great importance thanks to their own main characteristics [17]. HNTs are a naturally occurring alumino-silicate whose structural formula is

Al$_2$Si$_2$O$_5$(OH)$_4$·nH$_2$O, where Al is disposed in an gibbsite-like octahedral organization of Al–OH groups whereas Si–O groups form a tetrahedral sheet [18,19]. Both aluminols and siloxanes layers are overlapped in a kaolinite typical sheet that rolls up due to some structural defects and to the presence of water molecules, thus giving halloysite its peculiar narrow nanotubular structure [20–22].

HNTs dimensions depend on the natural deposit the clay is extracted from. In particular, the internal and external diameters are approximately 10–15 and 50–80 nm respectively, while the nanotubes length can range from 100 nm to 2 μm. [19] Interestingly, it is possible to classify halloysite by considering the distance between interlayers. For instance, it can be 7 or 10 Å depending on the number of water molecules present between the layers, which is namely 0 or 2, respectively [23,24]. Moreover, one of halloysite's most fascinating and important features is the different charge, in the pH interval from 3 to 8, between the outer surface that is mainly composed of Si–O groups and negatively charged, and the inner surface that is mainly composed of Al–OH and positively charged [25,26]. This different charge, due to the chemical composition, allows for selective functionalization, exploiting both the covalent and electrostatic interactions of each surface with other oppositely charged species: Drug molecules, polysaccharides, proteins, lipids, surfactants and so on [27–29]. All these features, and also considering that they are low cost, eco- and biocompatible materials [30], make HNTs suitable for designing hybrid materials for waste water remediation [31–35], cultural heritage treatment [36,37], biotechnological applications [38–44], and packaging [45–50].

Notably, halloysite is commonly used as a component in drug delivery systems through exploiting its characteristics in combination with other organic moieties, for example the temperature responsive polymers such as poly(N-isopropylacrylamide) (PNIPAAMs) that can selectively interact with the inner/outer surfaces thus influencing the release kinetics by changing their adsorption site [51], or natural occurring biopolymers for the preparation of end capped nanotubes with smart gates, or reverse inorganic micelles for the formation of nanohydrogels inside the HNTs lumen for a triggered absorption or release [52,53].

As evidenced in a recent review [17], the combination of polymer hydrogels and hollow inorganic nanotubes represents a perspective strategy for the fabrication of functional carriers in an advanced application.

In this work, we prepared hydrogel beads based on chitosan containing halloysite nanotubes. An alginate layer was introduced by diffusion and immersion of the beads in a sodium alginate solution. The dispersion of nanotubes into the hybrid gel and the localization of the alginate was investigated by SEM and fluorescence microscopy. Doxycycline, an antibiotic of the tetracycline class, was used as the model drug and it was loaded into the halloysite cavity by using a literature protocol [54]. This work represents a promising step for a valid alternative to generate hybrid hydrogels with oppositely charged polysaccharides and nanoclay with specific morphology for controlled drug release.

2. Results and Discussion

2.1. Morphology of the Hybrid Gel Beads

Figure 1 shows the optical image of the prepared chitosan/HNTs gel beads covered with a calcium alginate layer. They had an average diameter of 3.5 mm that shrinked to 0.8 mm when dried. To highlight the halloysite nanotubes distribution in the beads, SEM images were taken and they showed the presence of halloysite nanotubes in a random orientation within the polymer matrix (Figure 1). Similar findings have been reported by us for alginate/halloysite gel beads [55]. The dispersibility of the nanoparticles could be explained by the affinity between the polymer and halloysite and the colloidal stability of the nanoclay in polymer solution [56]. It should be noted that the halloysite cavity did not interact with the chitosan polycation as the inner surface of the nanotubes was positively charged [26]. Therefore, the HNTs lumen was preserved for drug loading.

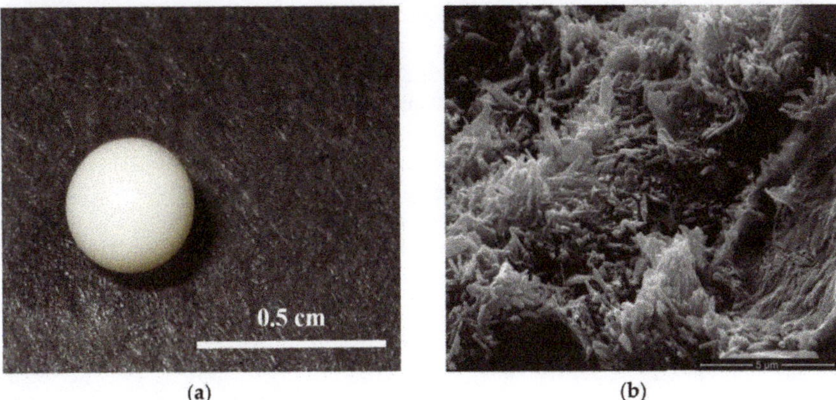

Figure 1. (a) Optical image of wet alginate/chitosan/halloysite nanotubes (HNTs) gel beads and (b) SEM image of the inner part of the dried gel beads.

Neither optical nor SEM imaging were able to identify the alginate location in the beads. We therefore thought to label the alginate polymer with a fluorescent probe 5-(4,6-dichlorotriazinyl) aminofluorescein (DTAF) that showed fluorescent emission when exited at 490 nm. Firstly, a blank experiment on chitosan/halloysite gel beads was carried out and negligible fluorescence was observed. The laser scanning confocal microscopy images on chitosan/HNTs gel beads covered with a calcium alginate layer clearly showed a fluorescent layer with an average thickness of approximately 130 μm, revealing that the diffusion of alginate into the chitosan/halloysite gel beads occurred up to a certain extent and the core of the beads was alginate free (Figure 2). On this basis, one could conclude that the simple preparation protocol allowed us to prepare a controlled complex architecture in mesoscopic scale that might be suitable for sustained release of active substances.

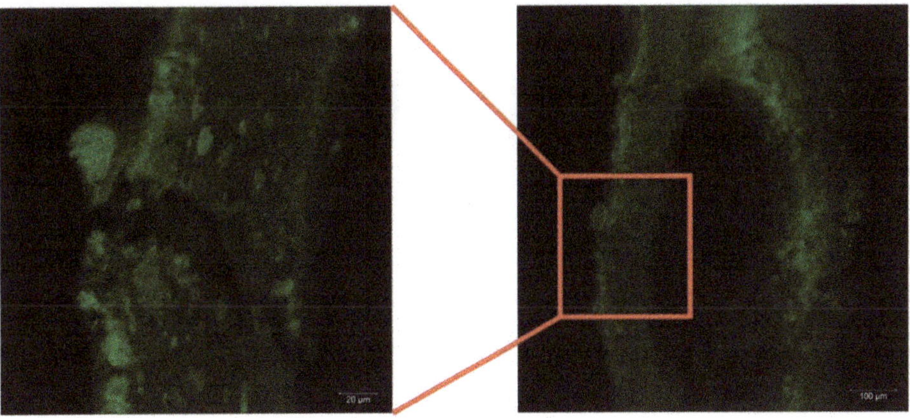

Figure 2. Laser scanning confocal microscopy images of alginate/chitosan/HNTs gel beads. Note that Alginate was labelled with 5-(4,6-dichlorotriazinyl) aminofluorescein (DTAF).

2.2. Drug Release Experiments

Release experiments were carried out by using doxycycline chlorohydrate as a model drug that could be loaded into the halloysite nanotubes [54]. The pK value for the drug was approximately 3 and the solubility dropped down as soon as the non-ionic form was obtained at pH > pK. Therefore, under our experimental conditions the drug was always neutral. The release profiles in water are provided in Figure 3 for doxycycline from halloysite nanotubes, chitosan/HNTs dried beads and alginate/chitosan/HNTs dried beads. It was clearly observed that halloysite incorporation into chitosan gel beds only slightly slowed down the drug release from being fully available in the solvent media in 20 min. The sustained release was due to the slow release of the drug from the nanotube cavity and the subsequent drug diffusion through the polymer matrix to the solvent. The presence of an alginate coating significantly slowed down the doxycycline release from the hybrid beads. In particular after 20 min only 50% of the drug was released into the solvent media while a full release occurs in more than 80 min. These results could be interpreted by considering that the alginate shell in combination with the oppositely charged chitosan could generate a highly viscous layer that further delayed the drug diffusion from the beads to the solvent.

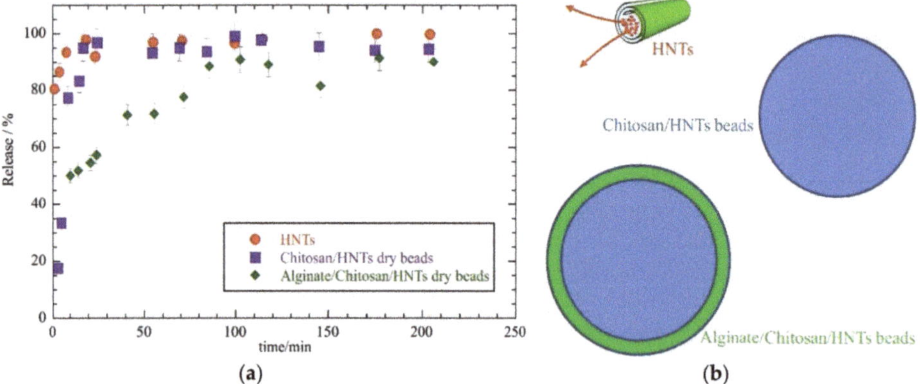

Figure 3. (**a**) Doxycycline chlorohydrate release as a function of time for different carriers. (**b**) Sketch view of the different release systems.

3. Materials and Methods

3.1. Materials

Halloysite, acetic acid, sodium hydroxide, ethylenediaminetetraacetic acid and DTAF, sodium alginate (M_w = 70–100 kg mol^{-1}), and chitosan (M_w = 50–190 kg mol^{-1}) were Sigma products. Doxycycline chlorohydrate ($C_{22}H_{24}N_2O_8 \bullet HCl$, M_w = 480.90 kg mol^{-1}) was from Alfa Aesar. Halloysite characterization was reported in our recent publication [19].

3.2. HNTs Loading with Doxycycline Chlorohydrate

The drug loading into HNTs cavity was carried out by using a procedure well eshtablished in literature [54]. Briefly, 0.5 g of doxycycline chlorohydrate was mixed with 2 g of HNTs in 20 cm^3 of water. The dispersion was kept under vacuum for 30 min. This procedure was repeated three times before centrifugation at 8000 rpm for 20 min to separate the loaded HNTs from the supernatant. The loaded HNTs were rinsed with water and the drug loading of 4.2 wt % was determined by thermogravimetry. The experimental thermogravimetric curves are provided in Supplementary Material. The method used for drug loading calculation was detailed elsewhere [57].

3.3. Preparation of Gel Beds

The chitosan based gel beads were prepared by using the dropping technique [58]. Chitosan (2 wt %) was dissolved in water containing 0.5 wt % of acetic acid. A peristaltic pump was used to drop the chitosan solution into an aqueous solution of NaOH 1.5 M. The needle diameter was 0.4 mm and the distance from the needle to the liquid surface was 2 cm. The obtained gel beads stood in the NaOH solution overnight, and afterwards they were rinsed with water three times. The preparation of the hybrid HNTs/Chitosan gel beads was carried out by using the same methodology. In this case, HNTs loaded with doxycycline chlorohydrate were dispersed into the chitosan solution with a polymer: HNTs weight ratio of 1:1. Some of the beads were in contact with a sodium alginate solution (2 wt %) for 10 mins and then with $CaCl_2$ 0.1 M to cross-link the alginate polymer. Beads were dried out at 40 °C overnight.

3.4. Doxycycline Chlorohydrate Release Experiments

The release profiles in water were determined by measuring UV-VIS spectra in a quartz cuvettes without stirring. In particular, one dried bead or the equivalent amount of loaded HNTs was weighted and directly placed into a cuvette. An amount of 2 cm^3 of distilled water was added and the spectra was recorded for 200 min.

3.5. Synthesis of DTAF Labeled Sodium Alginate

Alginate fluorescent labelling was carried out following the literature [59]. Sodium alginate (10 mg cm^{-3}) was solubilized in sodium bicarbonate (50 mM) and 1.0 M NaOH was used to adjust the pH to 9. DTAF (concentration of 10 mg mL^{-1} in dimethyl sulfoxide) was added at room temperature with an alginate: DTAF solutions volume ratio of 1:0.4. After one night of stirring, the mixture was dialysed in a 10 kDa cut-off dialysis tubing against phosphate-buffered saline (PBS) until the DTAF was not detected in the dialysate by UV at 490 nm.

3.6. Experimental Methods

UV-VIS spectra of doxycycline chlorohydrate were recorded by a Specord S600 (Analytik, Jena, Germany). Doxicycline chlorohydrate in water had a peak at 362 nm with an extinction coefficient of 23.6 ± 0.3 cm^3 mg^{-1}. SEM images were obtained by using a microscope ESEM FEI QUANTA 200F (FEI, Hillsboro, OR, USA) in high vacuum mode (<6 × 10^{-4} Pa). Before each SEM experiment, the surface of the sample was coated with gold in argon by means of an Edwards Sputter Coater S150A (Edwards Lifesciences, Milan, Italy) to avoid charging under an electron beam. Laser scanning confocal microscopy images were obtained using a LSM 780 instrument (Carl Zeiss, Jena, Germany) equipped with apochromatic 20× and 40× objectives and argon laser (488 nm). Images were processed using ZEN Black software (Carl Zeiss MicroImaging GmbH, Göttingen, Germany). Thermogravimetry (TG) measurements were performed by means of a Q5000 IR apparatus (TA Instruments, Milan, Italy) under nitrogen flows of 25 and 10 cm^3 min^{-1} for the sample and the balance, respectively. The sample (approximately 3 mg) was heated from room temperature to 700 °C at 10 °C min^{-1}. Calibration was carried out by following the procedure reported in literature [60].

4. Conclusions

In summary, we prepared hybrid gel beads with a chitosan rich core and an alginate rich shell containing halloysite nanotubes. The clay nanoparticles were loaded with a model drug and showed a good dispersion within the beads. The kinetics of the drug release was controlled by a core/shell structure. This work opens new perspectives into the preparation of hybrid biopolymer/nanoclay structures for drug delivery applications, and proposes a new strategy for obtaining tuned drug release.

Supplementary Materials: The following are available online at http://www.mdpi.com/2079-6412/9/2/70/s1, Figure S1: Thermogravimetric curves for HNTs loaded with Doxycycline chlorohydrate and their pure components (HNTs and Doxycycline chlorohydrate).

Author Contributions: Conceptualization, G.L. and R.F.; Methodology, G.C.; Validation, G.C., G.L. and R.F.; Formal Analysis, G.L.; Investigation, L.L. and F.P.; Data Curation, L.L.; Writing-Original Draft Preparation, L.L.; Writing-Review & Editing, G.L.; Supervision, G.L.; Project Administration, S.M.; Funding Acquisition, S.M. and R.F.

Acknowledgments: The work was financially supported by Progetto di ricerca e sviluppo "AGM for CuHe" (No. ARS01_00697) and the CORI 2018 project by the University of Palermo. The work was performed according to the Russian Government Program of Competitive Growth of the Kazan Federal University. This work was funded by the subsidy allocated to the Kazan Federal University for the state assignment in the sphere of scientific activities (No. 16.2822.2017/4.6) and by RFBR (No. 18-29-11031).

Conflicts of Interest: The authors declare no conflict of interest.

References

1. Ahmed, E.M. Hydrogel: Preparation, characterization, and applications: A review. *J. Adv. Res.* **2015**, *6*, 105–121. [CrossRef]
2. Akhtar, M.F.; Hanif, M.; Ranjha, N.M. Methods of synthesis of hydrogels: A review. *Saudi Pharm. J.* **2016**, *24*, 554–559. [CrossRef] [PubMed]
3. Ullah, F.; Othman, M.B.H.; Javed, F.; Ahmad, Z.; Akil, H.M. Classification, processing and application of hydrogels: A review. *Mater. Sci. Eng. C* **2015**, *57*, 414–433. [CrossRef] [PubMed]
4. Li, Y.; Huang, G.; Zhang, X.; Li, B.; Chen, Y.; Lu, T.; Lu, T.J.; Xu, F. Magnetic Hydrogels and Their Potential Biomedical Applications. *Adv. Funct. Mater.* **2012**, *23*, 660–672. [CrossRef]
5. Cao, J.; Tan, Y.; Che, Y.; Xin, H. Novel complex gel beads composed of hydrolyzed polyacrylamide and chitosan: An effective adsorbent for the removal of heavy metal from aqueous solution. *Bioresour. Technol.* **2010**, *101*, 2558–2561. [CrossRef] [PubMed]
6. Zheng, Y.; Wang, A. Enhanced Adsorption of Ammonium Using Hydrogel Composites Based on Chitosan and Halloysite. *J. Macromol. Sci. Part A* **2009**, *47*, 33–38. [CrossRef]
7. Li, J.; Li, X.; Zhou, Z.; Ni, X.; Leong, K.W. Formation of Supramolecular Hydrogels Induced by Inclusion Complexation between Pluronics and α-Cyclodextrin. *Macromolecules* **2001**, *34*, 7236–7237. [CrossRef]
8. Dadsetan, M.; Taylor, K.E.; Yong, C.; Bajzer, Z.; Lu, L.; Yaszemski, M.J. Controlled release of doxorubicin from pH-responsive microgels. *Acta Biomater.* **2013**, *9*, 5438–5446. [CrossRef]
9. Varaprasad, K.; Raghavendra, G.M.; Jayaramudu, T.; Yallapu, M.M.; Sadiku, R. A mini review on hydrogels classification and recent developments in miscellaneous applications. *Mater. Sci. Eng. C* **2017**, *79*, 958–971. [CrossRef]
10. Ganguly, K.; Chaturvedi, K.; More, U.A.; Nadagouda, M.N.; Aminabhavi, T.M. Polysaccharide-based micro/nanohydrogels for delivering macromolecular therapeutics. *Drug Deliv. Res. Asia Pac. Reg.* **2014**, *193*, 162–173. [CrossRef]
11. Wei, Z.; Yang, J.H.; Liu, Z.Q.; Xu, F.; Zhou, J.X.; Zrínyi, M.; Osada, Y.; Chen, Y.M. Novel biocompatible polysaccharide-based self-healing hydrogel. *Adv. Funct. Mater.* **2015**, *25*, 1352–1359. [CrossRef]
12. Sahiner, N.; Godbey, W.T.; McPherson, G.L.; John, V.T. Microgel, nanogel and hydrogel–hydrogel semi-IPN composites for biomedical applications: Synthesis and characterization. *Colloid Polym. Sci.* **2006**, *284*, 1121–1129. [CrossRef]
13. Liu, K.-H.; Liu, T.-Y.; Chen, S.-Y.; Liu, D.-M. Drug release behavior of chitosan–montmorillonite nanocomposite hydrogels following electrostimulation. *Acta Biomater.* **2008**, *4*, 1038–1045. [CrossRef]
14. Yang, H.; Hua, S.; Wang, W.; Wang, A. Composite Hydrogel Beads Based on Chitosan and Laponite: Preparation, Swelling, and Drug Release Behaviour. *Iran. Polym. J.* **2011**, *20*, 479–490.
15. Bonifacio, M.A.; Gentile, P.; Ferreira, A.M.; Cometa, S.; De Giglio, E. Insight into halloysite nanotubes-loaded gellan gum hydrogels for soft tissue engineering applications. *Carbohydr. Polym.* **2017**, *163*, 280–291. [CrossRef] [PubMed]
16. Lee, H.; Ryu, J.; Kim, D.; Joo, Y.; Lee, S.U.; Sohn, D. Preparation of an imogolite/poly(acrylic acid) hybrid gel. *J. Colloid Interface Sci.* **2013**, *406*, 165–171. [CrossRef] [PubMed]
17. Lazzara, G.; Cavallaro, G.; Panchal, A.; Fakhrullin, R.; Stavitskaya, A.; Vinokurov, V.; Lvov, Y. An assembly of organic-inorganic composites using halloysite clay nanotubes. *Curr. Opin. Colloid Interface Sci.* **2018**, *35*, 42–50. [CrossRef]
18. Lisuzzo, L.; Cavallaro, G.; Parisi, F.; Milioto, S.; Lazzara, G. Colloidal stability of halloysite clay nanotubes. *Ceram. Int.* **2018**. [CrossRef]
19. Cavallaro, G.; Chiappisi, L.; Pasbakhsh, P.; Gradzielski, M.; Lazzara, G. A structural comparison of halloysite nanotubes of different origin by Small-Angle Neutron Scattering (SANS) and Electric Birefringence. *Appl. Clay Sci.* **2018**, *160*, 71–80. [CrossRef]

20. Lvov, Y.M.; Shchukin, D.G.; Mohwald, H.; Price, R.R. Halloysite Clay Nanotubes for Controlled Release of Protective Agents. *ACS Nano* **2008**, *2*, 814–820. [CrossRef]
21. Luo, Z.; Song, H.; Feng, X.; Run, M.; Cui, H.; Wu, L.; Gao, J.; Wang, Z. Liquid Crystalline Phase Behavior and Sol–Gel Transition in Aqueous Halloysite Nanotube Dispersions. *Langmuir* **2013**, *29*, 12358–12366. [CrossRef] [PubMed]
22. Viseras, C.; Cerezo, P.; Sanchez, R.; Salcedo, I.; Aguzzi, C. Current challenges in clay minerals for drug delivery. *Appl. Clay Sci.* **2010**, *48*, 291–295. [CrossRef]
23. Joussein, E.; Petit, S.; Churchman, G.J.; Theng, B.; Righi, D.; Delvaux, B. Halloysite clay minerals—A review. *Clay Miner.* **2005**, *40*, 383–426. [CrossRef]
24. Pasbakhsh, P.; Churchman, G.J.; Keeling, J.L. Characterisation of properties of various halloysites relevant to their use as nanotubes and microfibre fillers. *Appl. Clay Sci.* **2013**, *74*, 47–57. [CrossRef]
25. Cavallaro, G.; Lazzara, G.; Milioto, S. Exploiting the Colloidal Stability and Solubilization Ability of Clay Nanotubes/Ionic Surfactant Hybrid Nanomaterials. *J. Phys. Chem. C* **2012**, *116*, 21932–21938. [CrossRef]
26. Bertolino, V.; Cavallaro, G.; Lazzara, G.; Milioto, S.; Parisi, F. Biopolymer-Targeted Adsorption onto Halloysite Nanotubes in Aqueous Media. *Langmuir* **2017**, *33*, 3317–3323. [CrossRef]
27. Abdullayev, E.; Price, R.; Shchukin, D.; Lvov, Y. Halloysite Tubes as Nanocontainers for Anticorrosion Coating with Benzotriazole. *ACS Appl. Mater. Interfaces* **2009**, *1*, 1437–1443. [CrossRef]
28. Cavallaro, G.; Lazzara, G.; Milioto, S.; Palmisano, G.; Parisi, F. Halloysite nanotube with fluorinated lumen: Non-foaming nanocontainer for storage and controlled release of oxygen in aqueous media. *J. Colloid Interface Sci.* **2014**, *417*, 66–71. [CrossRef]
29. Aguzzi, C.; Viseras, C.; Cerezo, P.; Salcedo, I.; Sánchez-Espejo, R.; Valenzuela, C. Release kinetics of 5-aminosalicylic acid from halloysite. *Colloids Surf. B Biointerfaces* **2013**, *105*, 75–80. [CrossRef]
30. Fakhrullina, G.I.; Akhatova, F.S.; Lvov, Y.M.; Fakhrullin, R.F. Toxicity of halloysite clay nanotubes in vivo: A Caenorhabditis elegans study. *Environ. Sci. Nano* **2015**, *2*, 54–59. [CrossRef]
31. Zhao, Y.; Abdullayev, E.; Vasiliev, A.; Lvov, Y. Halloysite nanotubule clay for efficient water purification. *J. Colloid Interface Sci.* **2013**, *406*, 121–129. [CrossRef] [PubMed]
32. Cavallaro, G.; Lazzara, G.; Milioto, S.; Parisi, F.; Sanzillo, V. Modified Halloysite Nanotubes: Nanoarchitectures for Enhancing the Capture of Oils from Vapor and Liquid Phases. *ACS Appl. Mater. Interfaces* **2014**, *6*, 606–612. [CrossRef]
33. Luo, P.; Zhang, J.; Zhang, B.; Wang, J.; Zhao, Y.; Liu, J. Preparation and Characterization of Silane Coupling Agent Modified Halloysite for Cr(VI) Removal. *Ind. Eng. Chem. Res.* **2011**, *50*, 10246–10252. [CrossRef]
34. Zhao, M.; Liu, P. Adsorption behavior of methylene blue on halloysite nanotubes. *Microporous Mesoporous Mater.* **2008**, *112*, 419–424. [CrossRef]
35. Hermawan, A.A.; Chang, J.W.; Pasbakhsh, P.; Hart, F.; Talei, A. Halloysite nanotubes as a fine grained material for heavy metal ions removal in tropical biofiltration systems. *Appl. Clay Sci.* **2018**, *160*, 106–115. [CrossRef]
36. Cavallaro, G.; Milioto, S.; Parisi, F.; Lazzara, G. Halloysite Nanotubes Loaded with Calcium Hydroxide: Alkaline Fillers for the Deacidification of Waterlogged Archeological Woods. *ACS Appl. Mater. Interfaces* **2018**, *10*, 27355–27364. [CrossRef] [PubMed]
37. Cavallaro, G.; Lazzara, G.; Milioto, S.; Parisi, F. Halloysite Nanotubes for Cleaning, Consolidation and Protection. *Chem. Rec.* **2018**, *18*, 940–949. [CrossRef]
38. Lvov, Y.; Abdullayev, E. Functional polymer–clay nanotube composites with sustained release of chemical agents. *Prog. Polym. Sci.* **2013**, *38*, 1690–1719. [CrossRef]
39. Abdullayev, E.; Sakakibara, K.; Okamoto, K.; Wei, W.; Ariga, K.; Lvov, Y. Natural Tubule Clay Template Synthesis of Silver Nanorods for Antibacterial Composite Coating. *ACS Appl. Mater. Interfaces* **2011**, *3*, 4040–4046. [CrossRef]
40. Liu, M.; Wu, C.; Jiao, Y.; Xiong, S.; Zhou, C. Chitosan-halloysite nanotubes nanocomposite scaffolds for tissue engineering. *J. Mater. Chem. B* **2013**, *1*, 2078–2089. [CrossRef]
41. Zhang, H.; Cheng, C.; Song, H.; Bai, L.; Cheng, Y.; Ba, X.; Wu, Y. A facile one-step grafting of polyphosphonium onto halloysite nanotubes initiated by Ce(iv). *Chem. Commun.* **2019**. [CrossRef] [PubMed]
42. Liu, F.; Bai, L.; Zhang, H.; Song, H.; Hu, L.; Wu, Y.; Ba, X. Smart H_2O_2-Responsive Drug Delivery System Made by Halloysite Nanotubes and Carbohydrate Polymers. *ACS Appl. Mater. Interfaces* **2017**, *9*, 31626–31633. [CrossRef] [PubMed]

43. Wei, W.; Minullina, R.; Abdullayev, E.; Fakhrullin, R.; Mills, D.; Lvov, Y. Enhanced efficiency of antiseptics with sustained release from clay nanotubes. *RSC Adv.* **2014**, *4*, 488–494. [CrossRef]

44. Kurczewska, J.; Cegłowski, M.; Messyasz, B.; Schroeder, G. Dendrimer-functionalized halloysite nanotubes for effective drug delivery. *Appl. Clay Sci.* **2018**, *153*, 134–143. [CrossRef]

45. Gorrasi, G.; Pantani, R.; Murariu, M.; Dubois, P. PLA/Halloysite Nanocomposite Films: Water Vapor Barrier Properties and Specific Key Characteristics. *Macromol. Mater. Eng.* **2014**, *299*, 104–115. [CrossRef]

46. De Silva, R.T.; Pasbakhsh, P.; Goh, K.L.; Chai, S.-P.; Ismail, H. Physico-chemical characterisation of chitosan/halloysite composite membranes. *Polym. Test.* **2013**, *32*, 265–271. [CrossRef]

47. He, Y.; Kong, W.; Wang, W.; Liu, T.; Liu, Y.; Gong, Q.; Gao, J. Modified natural halloysite/potato starch composite films. *Carbohydr. Polym.* **2012**, *87*, 2706–2711. [CrossRef]

48. Biddeci, G.; Cavallaro, G.; Di Blasi, F.; Lazzara, G.; Massaro, M.; Milioto, S.; Parisi, F.; Riela, S.; Spinelli, G. Halloysite nanotubes loaded with peppermint essential oil as filler for functional biopolymer film. *Carbohydr. Polym.* **2016**, *152*, 548–557. [CrossRef]

49. Sun, P.; Liu, G.; Lv, D.; Dong, X.; Wu, J.; Wang, D. Simultaneous improvement in strength, toughness, and thermal stability of epoxy/halloysite nanotubes composites by interfacial modification. *J. Appl. Polym. Sci.* **2016**, *133*. [CrossRef]

50. Kim, M.; Kim, S.; Kim, T.; Lee, D.K.; Seo, B.; Lim, C.-S. Mechanical and Thermal Properties of Epoxy Composites Containing Zirconium Oxide Impregnated Halloysite Nanotubes. *Coatings* **2017**, *7*, 231. [CrossRef]

51. Cavallaro, G.; Lazzara, G.; Lisuzzo, L.; Milioto, S.; Parisi, F. Selective adsorption of oppositely charged PNIPAAM on halloysite surfaces: A route to thermo-responsive nanocarriers. *Nanotechnology* **2018**, *29*, 325702. [CrossRef]

52. Cavallaro, G.; Danilushkina, A.A.; Evtugyn, V.G.; Lazzara, G.; Milioto, S.; Parisi, F.; Rozhina, E.V.; Fakhrullin, R.F. Halloysite Nanotubes: Controlled Access and Release by Smart Gates. *Nanomaterials* **2017**, *7*, 199. [CrossRef] [PubMed]

53. Cavallaro, G.; Lazzara, G.; Milioto, S.; Parisi, F.; Evtugyn, V.; Rozhina, E.; Fakhrullin, R. Nanohydrogel Formation within the Halloysite Lumen for Triggered and Sustained Release. *ACS Appl. Mater. Interfaces* **2018**, *10*, 8265–8273. [CrossRef] [PubMed]

54. Lvov, Y.M.; DeVilliers, M.M.; Fakhrullin, R.F. The application of halloysite tubule nanoclay in drug delivery. *Expert Opin. Drug Deliv.* **2016**, *13*, 977–986. [CrossRef]

55. Cavallaro, G.; Gianguzza, A.; Lazzara, G.; Milioto, S.; Piazzese, D. Alginate gel beads filled with halloysite nanotubes. *Appl. Clay Sci.* **2013**, *72*, 132–137. [CrossRef]

56. Lisuzzo, L.; Cavallaro, G.; Lazzara, G.; Milioto, S.; Parisi, F.; Stetsyshyn, Y. Stability of Halloysite, Imogolite, and Boron Nitride Nanotubes in Solvent Media. *Appl. Sci.* **2018**, *8*, 1068. [CrossRef]

57. Cavallaro, G.; Lazzara, G.; Milioto, S.; Parisi, F.; Ruisi, F. Nanocomposites based on esterified colophony and halloysite clay nanotubes as consolidants for waterlogged archaeological woods. *Cellulose* **2017**, *24*, 3367–3376. [CrossRef]

58. Kofuji, K.; Shibata, K.; Murata, Y.; Miyamoto, E.; Kawashima, S. Preparation and Drug Retention of Biodegradable Chitosan Gel Beads. *Chem. Pharm. Bull. (Tokyo)* **1999**, *47*, 1494–1496. [CrossRef]

59. Mackie, A.; Bajka, B.; Rigby, N. Roles for dietary fibre in the upper GI tract: The importance of viscosity. *Food Res. Int.* **2016**, *88*, 234–238. [CrossRef]

60. Blanco, I.; Abate, L.; Bottino, F.A.; Bottino, P. Thermal degradation of hepta cyclopentyl, mono phenyl-polyhedral oligomeric silsesquioxane (hcp-POSS)/polystyrene (PS) nanocomposites. *Polym. Degrad. Stabil.* **2012**, *97*, 849–855. [CrossRef]

 *coatings*

Article

Biocompatible Layers Obtained from Functionalized Iron Oxide Nanoparticles in Suspension

Daniela Predoi [1,*], Simona Liliana Iconaru [1], Mihai Valentin Predoi [2], Nicolas Buton [3], Christelle Megier [3] and Mikael Motelica-Heino [4]

[1] National Institute of Materials Physics, 405A Atomistilor Street, P.O. Box MG7, 077125 Magurele, Romania; simonaiconaru@gmail.com

[2] Department of Mechanics, University Politehnica of Bucharest, BN 002, 313 Splaiul Independentei, Sector 6, 10023 Bucharest, Romania; predoi@gmail.com

[3] HORIBA Jobin Yvon S.A.S., 6-18, Rue du Canal, 91165 Longjumeau CEDEX, France; nicolas.buton@horiba.com (N.B.); christelle.megier@horiba.com (C.M.)

[4] Institut des Sciences de la Terre D'Orleans (ISTO), UMR, 327, Centre National de la Recherche Scientifique CNRS Université d'Orléans, 1A rue de la Férollerie, CEDEX 2, 45071 Orléans, France; mikael.motelica@univ-orleans.fr

* Correspondence: dpredoi@gmail.com

Received: 2 November 2019; Accepted: 18 November 2019; Published: 20 November 2019

Abstract: Iron oxide nanoparticles have been extensively studied for challenges in applicable areas such as medicine, pharmacy, and the environment. The functionalization of iron oxide nanoparticles with dextran opens new prospects for application. Suspension characterization methods such as dynamic light scattering (DLS) and zeta potential (ZP) have allowed us to obtain information regarding the stability and hydrodynamic diameter of these suspended particles. For rigorous characterization of the suspension of dextran-coated iron oxide nanoparticles (D-MNPs), studies have been performed using ultrasound measurements. The results obtained from DLS and ZP studies were compared with those obtained from ultrasound measurements. The obtained results show a good stability of D-MNPs. A comparison between the D-MNP dimension obtained from transmission electron microscopy (TEM), X-ray diffraction (XRD), and DLS studies was also performed. A scanning electron spectroscopy (SEM) image of a surface D-MNP layer obtained from the stable suspension shows that the particles are spherical in shape. The topographies of the elemental maps of the D-MNP layer showed a uniform distribution of the constituent elements. The homogeneity of the layer was also observed. The morphology of the HeLa cells incubated for 24 and 48 h with the D-MNP suspension and D-MNP layers did not change relative to the morphology presented by the control cells. The cytotoxicity studies conducted at different time intervals have shown that a slight decrease in the HeLa cell viability after 48 h of incubation for both samples was observed.

Keywords: iron oxide; dextran; suspension stability; thin layer; cell viability

---

1. Introduction

In recent years, progress made in the area of materials science has opened great opportunities for the use of nanoparticles in biotechnological applications. Inorganic materials are one of the most investigated types of materials for their properties and offer tremendous opportunities in chemistry, physics and biology. These materials are considered promising candidates in the development of newly nanostructured materials and devices with controllable physicochemical and biological properties. Moreover, in recent decades, important advances have been made in the fabrication, characterization, and tailoring of the properties of nanoparticles, generating valuable solutions for numerous biomedical applications [1–3]. Among inorganic materials, magnetic particles have been intensively studied since their introduction in the 1970s for their exquisite properties for applications in medicine and biology for magnetic targeting (targeted drug delivery, targeted gene therapy, targeted radionuclides), magnetic resonance imaging (MRI), and immunoassay applications (gene therapy, cell labeling, ribonucleic acid (RNA) and deoxyribonucleic acid (DNA) purification, cell separation, cell purification), as well as hyperthermia generation [4–10]. Therefore, these materials cover the entire processes involved in current medical practices (i.e., diagnostics, therapy, and treatment). The use of small-scale iron oxide has been common practice approximately 40 years, but in recent decades a considerable deal of attention has been focused towards the study of several types of iron oxide nanoparticles, magnetite (Fe_3O_4), and maghemite (γ-Fe_2O_3), due to the special properties that they exhibit at the nanometric scale [1–10]. Magnetite is an iron oxide with a cubic inverse spinel in which oxygen forms a closed face-centered cubic (fcc) structure and Fe cations occupy the interstitial (tetrahedral and octahedral) sites. Due to this structure, the electrons can hop between the Fe^{2+} and Fe^{3+} ions in the octahedral sites, even at room temperature, making magnetite an important class of half-metallic materials [11]. Furthermore, using a convenient surface coating, magnetic nanoparticles can be dispersed using certain solvents and can be used to obtain homogenous suspensions, designated as ferrofluids [12–15]. In recent decades, biocompatible aqueous-based magnetic fluids have attracted a great deal of attention in the development of novel biomedical applications such as magnetic resonance imaging, magnetic targeted drug delivery, magnetic intracellular hyperthermia, magnetic cell separation, and alternating current (AC) magnetic field-assisted cancer therapy, etc. [2,12–17]. In order to obtain stable magnetite magnetic fluids, usually, a biocompatible surface coating, commonly a natural polymer, is used. Polysaccharides are common natural polymers, formed of repeated monosaccharides that have numerous reactive groups, such as hydroxyl (–OH) amino (–NH_2) groups. These materials are often used as coatings for magnetic nanoparticles due to their antifouling and biocompatible properties. Moreover, they are also generally appreciated due to their versatility for further modifications. Recent studies have highlighted that polysaccharides could be successfully used as anti-fouling coatings for magnetic resonance imaging agents in the reduction of protein adsorption and for improving the biocompatibility of magnetic nanoparticles [2,7,18–22]. Dextran is a natural glucose-based polysaccharide and is one of the most used polymers used in the coating of magnetic nanoparticles [18–20]. This natural polymer is preferred due to its exquisite properties of solubility, low toxicity, biocompatibility, and its high affinity to iron oxide nanoparticles [6,22]. The prevention of non-specific protein adsorption is of utmost importance in the development of biocompatible material agents for medical implants, diagnostics, and therapeutics [2,12–22]. In this context, preliminary studies regarding the development of magnetite magnetic fluids stabilized using dextran have been reported since 1978 by Syusaburo and Masalatsu [18], and later studied by different research teams (S19 and S20). Moreover, several nanoparticles of this class are now approved by the United States Food and Drug Administration (FDA) and are used successfully in the medical field. Feridex I.V. (ferumoxides) is the first nanoparticle-based iron oxide imaging agent that has been approved by FDA (it was approved in 1996) and it is used in the detection of liver lesions [7]. Since then, other versions, like Combidex (ferumoxtran-10) and Feraheme (ferumoxytol) have been approved to be used in human diagnostics and treatment. Due to their proven biocompatibility, dextran-coated iron oxide nanoparticles are already recognized for their use in the development of multifunctional imaging agents [19–21]. Recently, in order to improve their efficiency, researchers have focused on the development of thin layers of magnetic nanoparticles. Processing the material into a thin layer improves the treatment efficiency and allows the obtainment of a homogeneous coating with a controlled release of an active component, having the same resemblance as the starting material. In this context, the aim of this study was the obtainment of stable dextran-coated iron oxide nanoparticles solutions (D-MNPs) and using them in the development of uniform and homogenous thin layers for biomedical applications. Complementary studies such as dynamic light scattering (DLS), zeta potential (ZP), TEM, XRD and ultrasound measurements were performed on the D-MNP suspensions that were used to make the coatings. The morphology, uniformity, and chemical compositions of the obtained layers were also studied by scanning electron microscopy (SEM) and energy dispersive X-ray spectroscopy (EDS). Furthermore, cytotoxicity and cell viability studies were performed on both the D-MNP suspensions and the obtained layer.

2. Materials and Methods

2.1. Materials Reagents

Ferric chloride hexahidrate ($FeCl_3 \cdot 6H_2O$), ferrous chloride tetrahydrate ($FeCl_2 \cdot 4H_2O$), natrium hydroxide (NaOH), dextran ($H(C_6H_{10}O_5)_xOH$) with a molecular weight of ~40,000, perchloric acid ($HClO_4$), and hydrochloric acid (HCl) were acquired from Merck (Bucharest, Romania). Deionized water was also used. A silicon substrate, using the spin-coating procedure, was purchased from Sil'tronix Silicon Technologies (Archamps, France).

2.2. Synthesis of D-MNPs

Chloride hexahydrate ($FeCl_3 \cdot 6H_2O$) dissolved in water and ferrous chloride tetrahydrate ($FeCl_2 \cdot 4H_2O$) in 2 M HCl were mixed at room temperature. The molar ratio of $[Fe^{3+}]/[Fe^{2+}]$ was 2. The mixture was added drop by drop into the dextran solution (30% *w/v*) and 400 mL NaOH (2 M) solution at 90 °C under continuous stirring for about 45 min. The resulting solution was continuously stirred (200 rpm) for 1 h at 90 °C. A 5 M NaOH solution was added dropwise to obtain a pH equal to 12. The black precipitate was treated repeatedly with a 3 M perchloric acid solution (400 mL). The $[Fe^{2+}]/[Fe^{3+}]$ ratio in final suspension was about 0.05. The last separation was performed by centrifugation at 16,000 rpm for 1 h. Finally, the particles were re-dispersed into a dextran solution. Finally, the dextran-coated iron oxide nanoparticle (D-MNP) solutions were obtained.

2.3. Thin Layer of D-MNPs

To achieve the thin layer, 0.3 mL of the D-MNP resulting solution was used. Here, 0.3 mL was pipetted using a syringe on the top of the silicon substrate. During the deposition, the silicon substrate was rotated at 1000 rpm for 40 s. The process was repeated 20 times. After each coating, the layer was dried in a nitrogen atmosphere for 1 h at 70 °C immediately after coating. After the last coating, the resulting layer was dried in a nitrogen atmosphere for 1 h at 70 °C and then treated at 100 °C under vacuum for 1 h to further densify the film.

2.4. Characterization Methods

Ultrasonic measurements were performed in a 100 mL suspension of D-MNPs, which were previously stirred for 30 min at room temperature using a magnetic stirrer (Velp, Usmate, Italy). After stopping the stirring, ultrasound pulses were sent through D-MNP suspension [23]. The recording of digital signals from the oscilloscope at a very precise interval of 5.00 s was done using an electronic device. The evolution of the signals in time offers information both on the stability of the suspension and the attenuation of the signals in time. For signal processing, under the same experimental conditions, double distilled water was considered as the reference fluid.

Dynamic light scattering measurements (DLS) and ζ-potential evaluations were performed on a SZ-100 Nanoparticle Analyzer (25 ± 1 °C; laser wavelength 532 nm) from Horiba-Jobin Yvon (Horiba Ltd., Kyoto, Japan). The scattering angle was 173° and the primary data were obtained from the correlation function of the scattered intensity. The dispersion medium viscosity was 0.895 mPa·s. The obtained samples were diluted 10 times in ethanol before the investigations of the DLS and ζ-potential. To determine the hydrodynamic diameter (D_H), the sample was illuminated with a laser source that allowed the estimation of the particle diffusion velocity. For each analyzed sample, there were three recorded determinations. The final value was determined by averaging the three measurements.

In order to evaluate the morphology of the D-MNP sample, a scanning electron microscope (SEM) with a HITACHI S4500 (Hitachi Ltd., Chiyoda, Japan) was used. A dedicated attachment to the SEM apparatus allowed us to assess the chemical composition of the D-MNP sample by energy dispersive

X-ray spectroscopy (EDS) at 20 kV. The morphology of the samples was investigated by transmission electron microscopy using a CM 20 (Philips-FEI, Hillsboro, OR, USA) transmission electron microscope equipped with a Lab6 filament (Agar Scientific Ltd., Stansted, UK) operating at 200 kV. The XRD patterns of the synthesized D-MNP samples were recorded with a Bruker D8 Advance diffractometer (Bruker, Karlsruhe, Germany), using Cu Kα (λ = 1.5418 Å) radiation in the 2θ ranging from 20° to 70° and a high efficiency LynxEye™ linear detector (Bruker, Karlsruhe, Germany). The crystallite sizes of the D-MNP samples were appraised from the X-ray line broadening using Scherrer formula:

$$d = K\lambda/\beta\cos\theta$$

where d is the mean size of the ordered (crystalline) domains, which may be smaller or equal to the grain size, K is a dimensionless shape factor, with a value close to unity. The shape factor has a typical value of about 0.9, but varies with the actual shape of the crystallite, λ is the X-ray wavelength, β is the line broadening at half the maximum intensity (FWHM) after subtracting the instrumental line broadening, in radians, and θ is the Bragg angle.

2.5. Hela Cell Viability Assays

The cytotoxicity assays of the samples were performed using the HeLa cell line Sigma-Aldrich Corp., St. Louis, MO, USA). In order to study the in vitro interaction of the HeLa cells with the samples, two-time intervals were chosen (24 and 48 h). The HeLa cells were trypsinized, counted, and seeded into 24-well plates at a density of 5×10^4 cells per well. The plates were incubated in Dulbecco's modified Eagle's medium (DMEM) (Sigma-Aldrich Corp., St. Louis, MO, USA), supplemented with 10% fetal bovine serum (Sigma-Aldrich Corp., St. Louis, MO, USA), at 37 °C, in a humid atmosphere of 5% CO_2 for 24 h. To assess the cytotoxicity of the D-MNP solutions, the cells were treated with 100 µg/mL of solution. The treated cells were then kept for 24 and 48 h at 37 °C, in a 5% CO_2 atmosphere, and a quantitative assay of the HeLa cell's viability was conducted using the conventional 3-(4,5-dimethylthiazolyl-2)-2,5-diphenyltetrazolium bromide (MTT) reduction assay. The results were interpreted as percentages of the viable cells treated with D-MNP solution relative to the viability of untreated cells (control, 100%). In the case of the D-MNP layers, the HeLa cells previously cultured in Dulbecco's modified Eagle's medium (DMEM) (Sigma-Aldrich Corp., St. Louis, MO, USA), supplemented with 10% fetal bovine serum (Sigma-Aldrich Corp., St. Louis, MO, USA), at 37 °C, in a humid atmosphere of 5% CO_2 for 24 h, were incubated with the D-MNP layers for 24 and 48 h. After 24 and 48 h of incubation, the cell viability was evaluated using the MTT reduction assay. Furthermore, the morphology and the cell cycle features of the HeLa cells incubated with both D-MNP suspensions and D-MNP layers were also analyzed. For that purpose, after being incubated for 24, respective to 48 h with the HeLa cell cultures, the samples were fixed using cold ethanol, stained with propidium iodide (PI), and visualized with the aid of a Leica DFC 450C fluorescence microscope (Leica Camera, Wetzlar, Germany). After growing HeLa cells in the presence of the samples for 24 and 48 h, the cell suspension was fixed in 70% ethanol for at least 30 min. After centrifugation, the cell pellet was resuspended in a 1 mL phosphate buffer containing 50 g/mL RNA-z and PI. The stained cells were analyzed at the flow cytometer within 2 h after fixation. DNA histogram deconvolution analysis was performed using FlowJo software (FlowJo v9) and the Watson model, measuring the percentage of cells in the G0/G, S, G2/M phases. The cytotoxic assays were conducted in triplicate and the results were presented as mean ± SD.

2.6. Statistical Analysis

The biological studies were carried out in triplicate. To perform the statistical analysis, a t-test and analysis of variance (ANOVA) were used. Differences between samples were considered to be significant at $p < 0.05$.

3. Results and Discussions

To estimate the stability of the D-MNP suspension used to obtain the thin layers, a non-destructive ultrasound-based technique was used.

The ultrasound studies were performed on the D-MNP suspension obtained by the coprecipitation method. The stability of the D-MNP suspension was studied both according to the evolution in time of the signals obtained from sending ultrasonic pulses through the studied suspension, depending on the evolution in time of the attenuation. The processing of the obtained results was performed according to the results obtained for double distilled water under the same experimental conditions. The exact determination of the speed of ultrasound in the D-MNP suspension for each signal was established according to the time delays between the first three recorded echoes and those of the reference suspension (bi-distilled water in our case). For the D-MNP suspension, the calculated velocity at 24 °C was 1508.2 m/s, while for the reference fluid it was 1494.84 m/s. This speed cannot be a reliable parameter in the characterization of the evolution of the sample as it has very little variation during the sedimentation of the particles. From the maximum amplitude of the transmitted signals compared to the recording moments, a more significant variation was determined. As can be seen in Figure 1, there was a slow and continuous variation of amplitudes, which will be used for further processing. Moreover, during the recording (3200 s), the suspension did not precipitate. On the other hand, the commonly encountered passage of the separation surface in front of the transducers does not appear in this case. As can be seen in the figure, there is a very brief initial interval of less than 60 s, during which the amplitude increases faster, followed by more than 3000 s of slow evolution. The first transmitted echo, which is best recorded, tends slowly towards a relative amplitude which is below 1, the value which would correspond to pure water. This remark is also an indicator of the high stability of the suspension, which will be quantitatively determined in the following section.

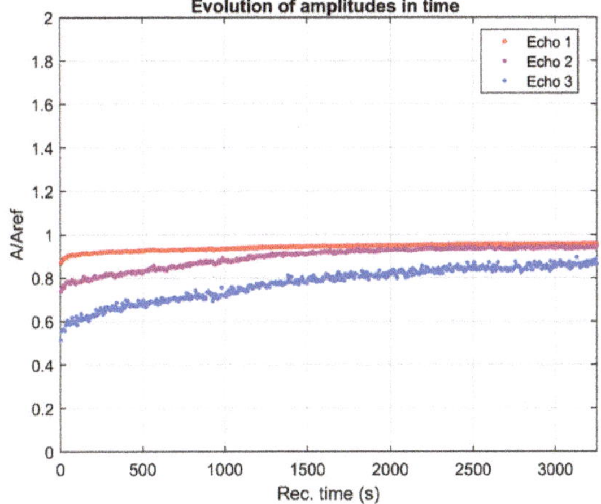

Figure 1. Relative amplitudes evolution vs. the recording moments.

The entire measurement period is relevant for the stability of this particular suspension. It can be observed that during this period of time, the amplitude of the first echo, which was determined with the highest precision, showed a slowly increasing value. The slope of this amplitude vs. time is related to the stability parameter, which in this case is $s = \frac{1}{A_m}\left|\frac{dA}{dt}\right| = 0.0000444$ (1/s), with A_m being the averaged amplitude of the signals. This small value indicates an excellent stability. It is reminded here that for pure water $s = 0$. The suspension has a typical signature in the frequency spectrum of the first transmitted echo (Figure 2).

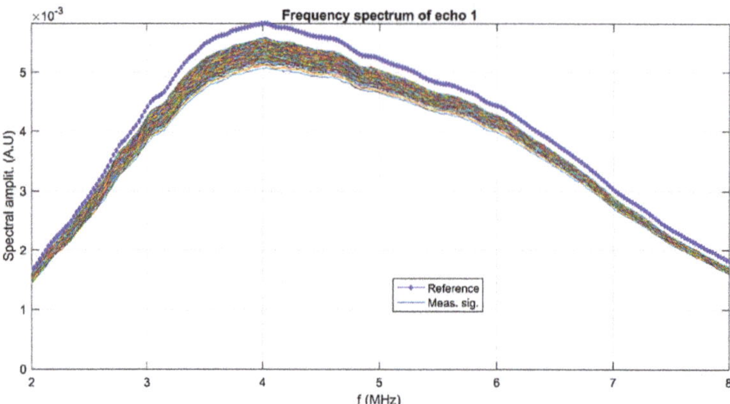

Figure 2. Frequency spectrum of the first transmitted echo. Reference fluid (◊) and all recorded echoes superimposed (many colors).

The lowest curves correspond to the first period of sedimentation and the upper ones correspond to the asymptotic sedimentation of the suspension fluid. The peak is at 4 MHz, which is a characteristic of the transducers. The spectrum for the reference fluid is indicated by blue markers. Another piece of information concerning the suspension is the spectral amplitude variation during the experiment. The recorded first echo was split into harmonic components of frequencies between 2 and 8 MHz (Figure 3). It is interesting to point out that the component at 2 MHz tends, on average, to a relative amplitude of 1, meaning that the suspension becomes transparent to signals at this frequency. All higher frequencies exhibit a continuous increasing path for their relative amplitudes, from a minimum of 0.86 towards 0.97 during the long recording interval. The evolution for each frequency is difficult to extract from Figure 3.

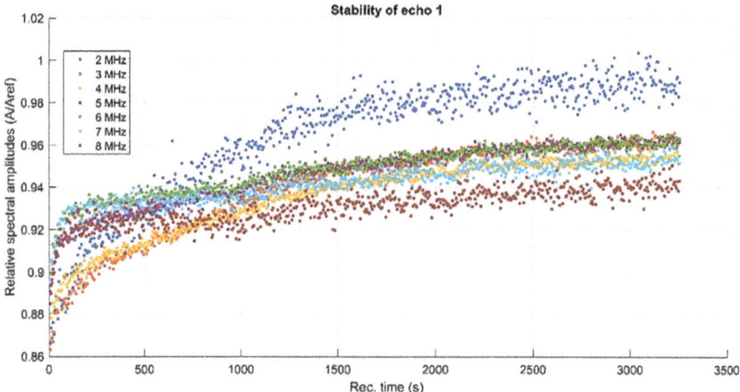

Figure 3. Spectral amplitudes relative variation vs. time for the first echo.

More explicit information on the attenuation dependency of frequency is given in Figure 4a, compared to the same dependency on frequency for the reference fluid. The attenuation is considerably higher (1.3–3.6 nepper/m) for the sample compared to 0.3–1.3 nepper/m for the reference fluid. A characteristic of this suspension is the local maxima of attenuation between 3 and 4 MHz, at which acoustic energy is absorbed and scattered more than in an equivalent homogeneous fluid.

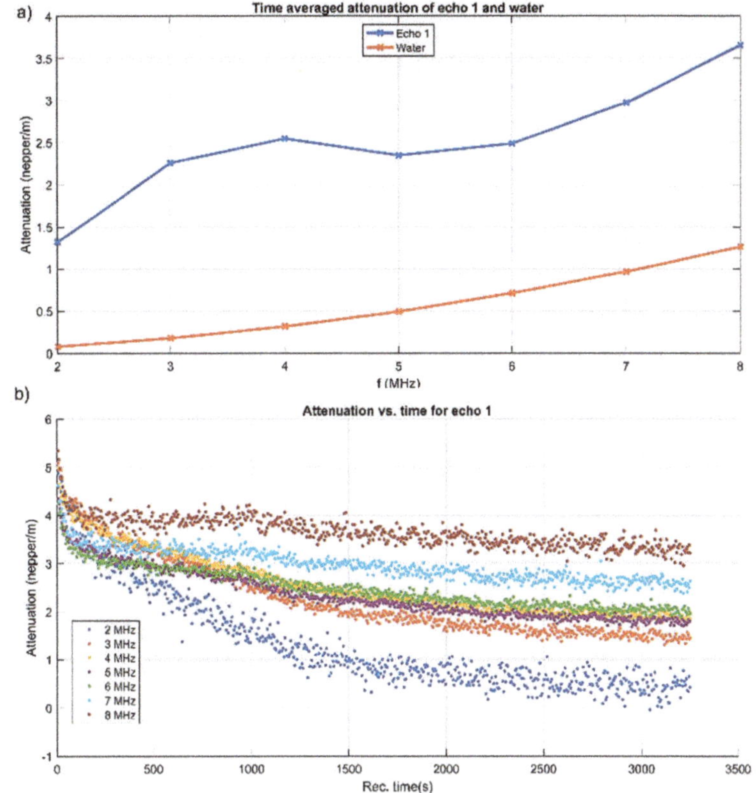

Figure 4. Attenuation vs. frequency for the first transmitted echo (**a**) and attenuation vs. time for the spectral components of echo 1 (**b**).

Certainly, the attenuation for each spectral component depends on the moment during the experiment, as shown in Figure 4b. At all selected frequency components, the initial attenuation was in a narrow interval of 4–5.5 nepper/m. The components at 2 and 3 MHz kept a higher attenuation in time, which descended towards 2.6 and respectively 3.3 dB/m after 3250 s of monitoring time. The other frequencies diminished their attenuation to lower than 2 nepper/m. Only the 2 MHz component tended to a very low attenuation. As a partial conclusion, a transducer with a central frequency as low as 2 MHz would not be capable to extract the distinct features of this sample.

The D-MNP suspension reaches a very stable state after about 1 min and remains stable, with a maximum amplitude variation of 4%, for a very long time.

Complementary studies to investigate the stability of the suspension to be used in the production of thin layers were performed by DLS. DLS and ζ-potential studies were performed on diluted solutions while ultrasound measurements were performed on the solution used to make the thin layers.

The DLS and ζ-potential investigations were performed on dilutions of the stable suspension of D-MNPs maintained under ambient conditions. The results obtained from the DLS investigations and ζ-potential are presented in Figure 5. For the DLS and ζ-potential measurements, the suspensions of D-MNP were diluted 5 times. The average hydrodynamic size (D_H) obtained by DLS was 27.8 ± 5 nm (Figure 5a). On the other hand, it was observed that the particles were monodispersed.

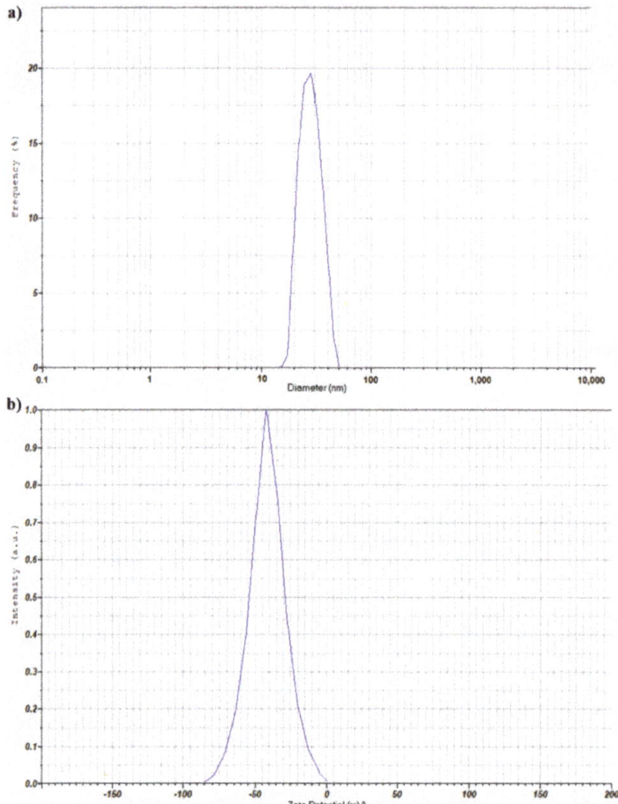

Figure 5. Particle size distribution (**a**) from DLS measurement and ζ-potential curves (**b**) of the dextran-coated iron oxide nanoparticle (D-MNP) suspensions.

The stability of the D-NMPs suspension depends on the electrostatic repulsion. As a result, measuring the ζ-potential for D-MNP suspensions is very important. Here, the ζ-potential value for D-NMs suspensions was −44.1 mV (Figure 5b). This value of the ζ-potential ensures an electrostatic repulsion force between particles that is large enough to prevent the attraction and collision caused by Brownian motion. The value of the ζ-potential showed that the stability of the D-MNP suspension was good.

The morphology and size of D-MNPs was evaluated by TEM and SEM analysis (Figure 6). The TEM results showed that dextran-coated iron oxide particles are spherical in shape and are dispersed (Figure 6a). The average diameter calculated from the size distribution was 7.3 ± 0.6 nm (Figure 6b). The size distribution was obtained after measuring about 1000 particles. A SEM micrograph of the D-MNP particles is presented in Figure 6c. It can see that the morphology of the D-MNP particles was uniform. The particles are spherical, with nanoscale dimensions. The average diameter calculated from SEM image was 8.5 ± 0.4 nm (Figure 6d).

In order to investigate the crystalline structure of D-MNPs, the D-MNPs suspensions were centrifuged at 12,000 rot/min and the resulting powder was dried in an oven at 80 °C. The X-ray diffractogram of the D-MNP powder after drying at 80 °C is presented in Figure 6e. All diffraction characteristic peaks corresponding to (220), (311), (400), (511) and (440) planes were identified. The identified diffraction peaks were indexed to a face-centered cubic spinel structure (Fd3m3 space group) with a lattice parameter of $a = 8.334$ Å (JCPDS card No. 39-1346). The lattice parameter of the synthesized

D-MNP (a = 8.334 Å) was in agreement with the bulk lattice parameter of maghemite (a = 8.3474 Å). These results were in agreement with those previously reported in the literature [8,24–26]. Scherrer's equation [27] was used to calculate the crystallite size of D-MNPs from the XRD line broadening. The calculated crystallite size (D_{XRD}) was 7.1 ± 0.1 nm. The very good crystallinity of the D-MNPs was confirmed by the XRD patterns.

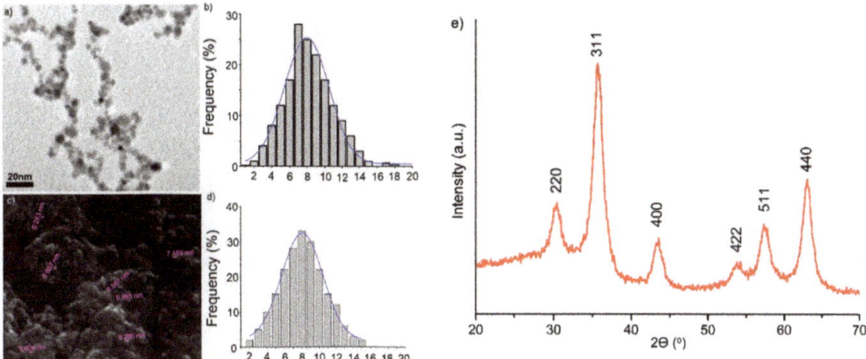

Figure 6. Large-area TEM (**a**), size distributions from TEM (**b**), SEM image (**c**) and size distributions from SEM (**d**) of iron oxide coated with dextran and X-ray diffraction pattern of dextran-coated maghemite nanoparticle (D-MNPs) (**e**).

The D_{XRD} was in good agreement with the size of D-MNPs obtained from TEM micrographs (D_{TEM}). On the other hand, the D_H obtained by DLS was substantially higher than the D_{TEM} obtained from the TEM micrographs. This difference may be due to the fact that the TEM analysis was done on powder while the DLS studies were performed on suspensions. The difference between the TEM and DLS results may be due to the fact that TEM analysis cannot measure any particle coating agent, while DLS provides information on both the particle diameter and the molecules or ions that are attached to its surface [28]. On the other hand, the DLS technique also measured possible aggregates of particles in the solution. By observing the molecules (dextran in our case) attached to the surface of the particles, the diameter determined by DLS measurements will be larger than that observed by TEM [29]. According to previous studies [30] it is possible that the particles may remain well dispersed/stable by coating with a polymer (in our case, dextran). It is known that when the particles are uncoated they tend to become agglomerated due to strong interparticle interactions.

The morphology and thin surface layers obtained from the stable suspension of D-MNPs were evaluated by SEM investigations (Figure 7a). Here, a SEM image of the surface D-MNP layer shows that the particles are spherical in shape. For a precise morphology assessment of the D-MNP layer, the 2D micrograph has been converted into a 3D surface map image (Figure 7b) using specialized software [31–33]. The chemical composition of the thin D-MNP layer was estimated by an Energy-dispersive X-ray spectroscopy (EDX) survey (Figure 7c). The peaks of the typical chemical element constituents of the D-MNPs were identified (i.e., C, O, and Fe). No other peaks that could belong to impurities were observed. This result shows the good purity of the obtained D-MNPs. In the topographies of the elemental maps of the D-MNP layer, a homogeneous and uniform distribution of the constituent elements was revealed.

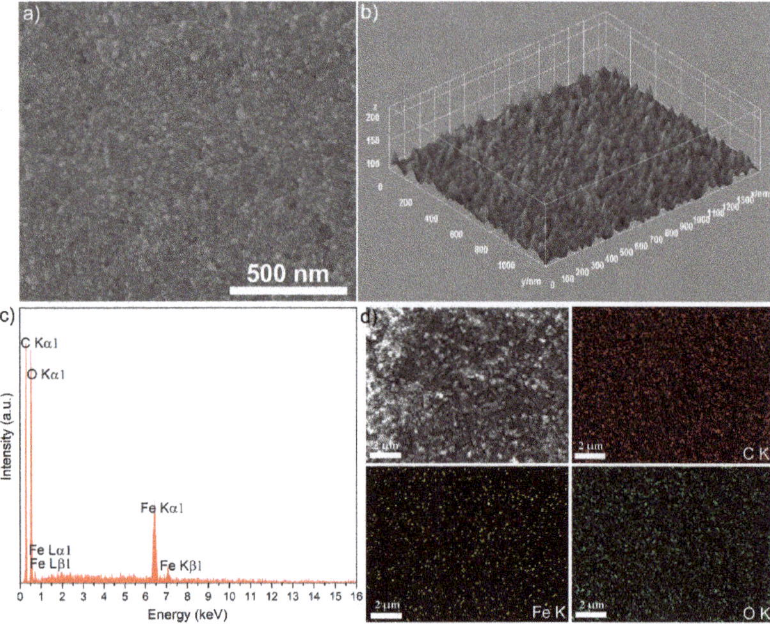

Figure 7. SEM image of surface the D-MNP layer: 2D (**a**), 3D (**b**), and the energy-dispersive X-ray spectroscopy spectra of D-MNP layer (**c**) and the elemental mapping of the D-MNP layer (**d**).

The toxicity of the D-MNP suspension and D-MNP layers were investigated using the first immortal human cell line (HeLa cells), which is known to be a remarkably durable and prolific cell line [34]. The HeLa cells were incubated for 24 and 48 h with the D-MNP suspension and D-MNP layers, then visualized using a fluorescence microscope. The morphology of the HeLa cells treated with the D-MNP suspension and D-MNP layers after 24 and 48 h of incubation are presented in Figure 8a–e. Figure 8b,c depicts the morphology of the HeLa cells incubated for 24 h with the investigated samples. The morphology of the HeLa cells incubated for 48 h with the samples are depicted in Figure 8e,f. The morphology of the HeLa cells grown in the culture medium for 24 and 48 h used as control are depicted in Figure 8a,d. The results of the fluorescence microscopy revealed that the morphology of the HeLa cells incubated for 24 and 48 h with the D-MNP suspension and D-MNP layers did not change relative to the morphology presented by the control cells. Furthermore, the influence of the D-MNP suspension and D-MNP layers on the HeLa cell cycle was also investigated after 24 and 48 h of incubation by flow cytometry. The results of the cell cycle analysis are presented in Figure 8g–l. The DNA histogram deconvolution was performed using the FlowJo software (FlowJo v9). The percentage of cells in the G0/G, S, and G2/M phases were quantified using the Watson model. It was observed that all tested compounds exhibited a small-scale level of cytotoxicity against HeLa cell cycle development for all of the tested time intervals. The flow cytometry histograms, depicted in Figure 8g–l, are in good agreement with the microscopy images of the HeLa cell culture incubated for 24 and 48 h with D-MNP suspension and D-MNP layers. The cell cycle analysis demonstrated that both the D-MNP suspension and D-MNP layers did not present a significant toxicity towards HeLa cells, even after 48 h of incubation.

The cell cycle diagrams of the HeLa cells incubated for 24 and 48 h with the D-MNP suspension (Figure 8h,k) and D-MNP layers (Figure 8i,l) were comparable to the cell cycle diagrams obtained for the HeLa cells grown in culture medium used as control (Figure 8g,j). The results of the percentage of cells in the G0/G, S, G2/M phases quantified using the Watson model revealed that there was no blockage of the eukaryotic G2/M phase, indicating no toxicity.

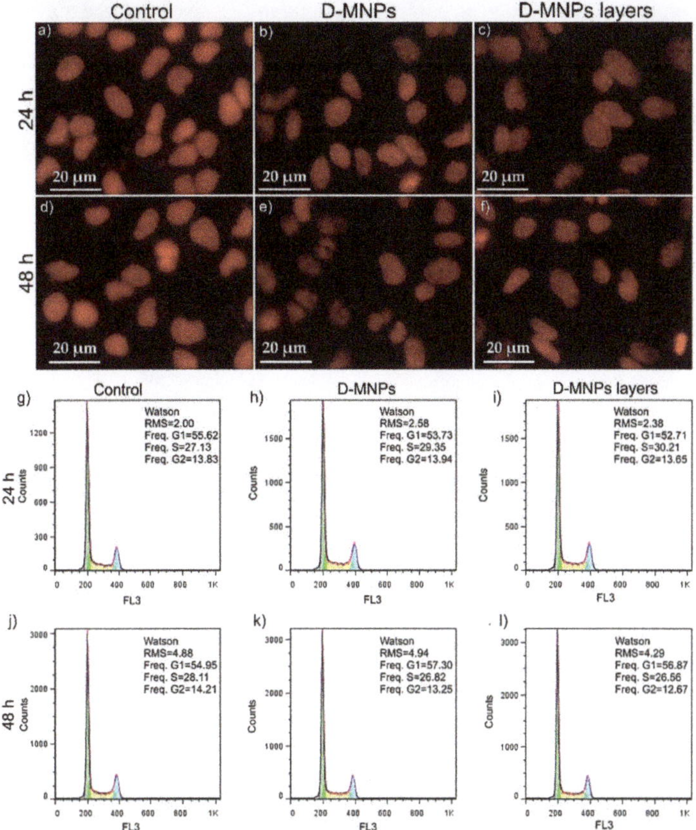

Figure 8. The morphology of the HeLa cells incubated with the D-MNP suspension (**b**) and D-MNP layers (**c**) at 24 h, relative to the control (**a**). The morphology of the HeLa cells incubated with the D-MNP suspension (**e**) and D-MNP layers (**f**) at 48 h, relative to the control (**d**). The cell cycle histogram analysis of HeLa cells incubated with D-MNP suspension (**h**) and D-MNP layers (**i**) at 24 h, relative to the control (**g**). Cell cycle analysis of the HeLa cells incubated with the D-MNP suspension (**k**) and D-MNP layers (**l**) at 48 h, relative to the control (**j**).

In order to assess the cytotoxicity of the D-MNP suspension and D-MNP layers, a quantitative MTT assay was performed to determine the cell viability of the HeLa cells after the 24 and 48 h incubation time intervals, in the presence of the D-MNP suspension and D-MNP layers. The results of the cytotoxicity tests at different incubation time periods (24 and 48 h) of the D-MNP suspension and D-MNP layers on the HeLa cell lines are shown in Figure 9. The MTT studies revealed that after 24 h of incubation time, there were no representative differences in the values of cells viabilities between the cells treated with D-MNPs suspension and D-MNPs layers and the control cell culture. The results have emphasized that after 24 h of incubation, the cell viability of HeLa cells was 96% in the case of the D-MNP suspension and 92% in the case of the D-MNP layers. Moreover, the MTT assay results show that after 48 h of incubation a slight decrease in HeLa cell viability was observed for both samples. In this case, a cell viability of 92% was obtained in the case of HeLa cells exposed to the D-MNP suspension and a cell viability of 88% was obtained in the case of HeLa cells exposed to the D-MNP layers. The results highlight that there is a correlation between the incubation time and the small toxicity effects presented by the tested samples. Moreover, the MTT suggested that the D-MNP suspension had better biocompatible properties than the D-MNP layers. The results obtained in the present study are in good agreement with previous studies regarding the toxicity of iron oxide nanoparticles [35–42]. Moreover, the results obtained from the statistical analysis, where $p = 0.042$, demonstrated that the observed difference observed in the cytotoxicity assays is unlikely to be due to chance, indicating a significant finding.

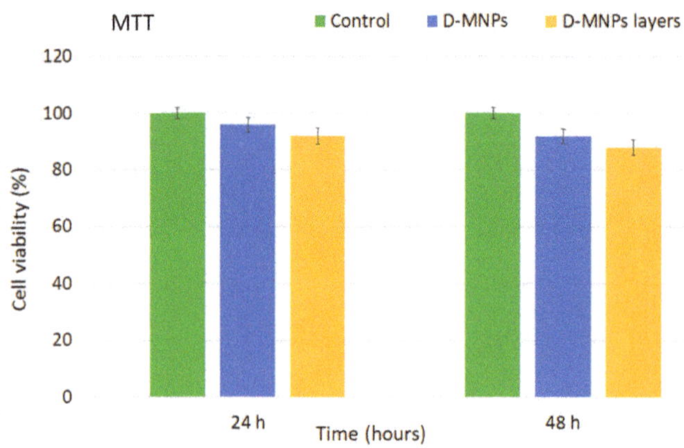

Figure 9. HeLa cell viability after incubation with the D-MNP suspension and the D-MNP layers at different incubation times.

Usually, the cytotoxicity induced by nanomaterials and nanolayers depends on numerous parameters, such as the type of the nanoparticles, their size and shape, the concentration, incubation time, and also the type of cell line [35–38]. In their study regarding, the "synthesis, characterization and toxicological evaluation of iron oxide nanoparticles in human lung alveolar epithelial cells", Dwivedi et al. [39] reported that the cell viability of A-549 cells diminished to 84%, 72%, and 56% after being incubated for 24 h with different concentrations, specifically, 10, 25, and 50 µg/mL, respectively, of iron oxide nanoparticles (IONPs). Another study conducted by Karlsson et al. [40] regarding the cytotoxicity of several metal oxide nanoparticles reported that iron oxide particles (Fe_3O_4, Fe_2O_3) exhibited none or low toxicity, while $CuZnFe_2O_4$ particles induced DNA damage to the A549 cell line. During recent years, several materials have been employed as coatings for IONPs in order to stabilize their physicochemical and biological properties. Villanueva et al. [41] studied "the influence of surface functionalization on the enhanced internalization of magnetic nanoparticles in cancer cells", and their results emphasize that iron oxide nanoparticles coated with dextranamino at concentrations of 0.05, 0.1, and 0.5 mg/mL present a low toxicity against HeLa cells. Moreover, Ankamwar et al. [42] demonstrated in their study that IONPs coated with a bipolar surfactant, namely, tetramethylammonium 11-aminoundecanoate, at concentrations of 0.1 to 10 µg/mL had no toxicity against HeLa cells. Furthermore, they have reported that the cytotoxicity of IONPs coated with a bipolar surfactant, tetramethylammonium 11-aminoundecanoate, was strongly dependent on the nanoparticle concentration [42].

4. Conclusions

The purpose of this study was to obtain stable suspensions in order to achieve homogenous and uniform coatings. The results obtained by complementary analysis techniques revealed the good stability of the suspension. The surface of the realized layer was uniform and homogeneous, with no cracks. Moreover, the uniform distribution of the constituent elements (C, O, and Fe) were also observed in the topographies of the elemental maps of the D-MNP layers.

The biocompatibility of the D-MNP suspension and D-MNPs layers were investigated using the HeLa cell line after 24 and 48 h of incubation. The qualitative cytotoxicity assays, using fluorescence microscopy, revealed that the D-MNP suspension and D-MNP layers did not present any toxicity towards the HeLa cells after 24 and 48 h of incubation. Moreover, the analysis of the cell cycle histogram demonstrated that both the D-MNP suspension and D-MNP layers had no significant toxic effects against the development of HeLa cells. Furthermore, the quantitative MTT assays proved that the D-MNP suspension and D-MNP layers had a negligible toxic effect on the development of the HeLa cells and that the effect was dependent on the incubation time. Therefore, both D-MNP suspensions and D-MNP layers are good candidates for use in the development of biomedical and cancer research devices.

Author Contributions: Conceptualization, D.P. and M.V.P.; Methodology, D.P., M.V.P. and M.M.-H.; Software, M.V.P.; Validation, D.P., S.L.I., M.M.-H. and M.V.P.; Formal analysis, D.P., S.L.I., M.V.P., M.M.-H., N.B. and C.M.; Investigation, D.P., S.L.I., M.V.P., M.M.-H., N.B. and C.M.; Resources, D.P., S.L.I., M.V.P., M.M.-H., N.B. and C.M.; Data curation, D.P., S.L.I. and M.V.P.; Writing—Original Draft Preparation, D.P., S.L.I. and M.V.P.; Writing—Review and Editing, D.P., S.L.I., M.V.P., M.M.-H., N.B. and C.M.; Visualization, D.P., S.L.I., M.V.P., M.M.-H., N.B. and C.M.; Supervision, M.V.P., M.M.-H., N.B. and C.M.; Project Administration, D.P.; Funding Acquisition, D.P.

Funding: This research was partially funded by a grant of the Romanian Ministry of Research and Innovation (PCCDI-UEFISCDI, project number PN-III-P1-1.2-PCCDI-2017-0629/contract No. 43PCCDI/2018 and PN-III-P1-1.2-PCCDI-2017-0134, Contract No. 23PCCDI/2018.

Acknowledgments: We want to thank Annie Richard and Audrey Sauldubois from the "Centre de Microscopie Electronique" of University of Orléans for assistance in SEM and TEM data acquisition, Alina Mihaela Prodan and Monica Liliana Badea for assistance with the in vitro experiments.

Conflicts of Interest: The authors declare no conflict of interest.

References

1. Curtis, A.S.G.; Wilkinson, C. Nanotechniques and approaches in biotechnology. *Trends Biotechnol.* **2001**, *19*, 97–101. [CrossRef]

2. Gupta, A.K.; Gupta, M. Synthesis and surface engineering of iron oxide nanoparticles for biomedical applications. *Biomaterials* **2005**, *26*, 3995–4021. [CrossRef] [PubMed]

3. Dragu, D.L.; Necula, L.G.; Bleotu, C.; Diaconu, C.C.; Chivu-Economescu, M. Therapies targeting cancer stem cells: Current trends and future challenges. *World J. Stem Cells* **2015**, *7*, 1185–1201. [PubMed]

4. Lu, A.H.; Salabas, E.L.; Schuth, F. Magnetic nanoparticles: Synthesis, protection, functionalization, and application. *Angew. Chem. Int.* **2007**, *46*, 1222–1244. [CrossRef] [PubMed]

5. Pankhurst, Q.A.; Connolly, J.; Jones, S.K.; Dobson, J. Applications of magnetic nanoparticles in biomedicine. *J. Phys. D Appl. Phys.* **2003**, *36*, R167–R181. [CrossRef]

6. Barbosa-Barros, L.; García-Jimeno, S.; Estelrich, J. Formation and characterization of biobased magnetic nanoparticles double coated with dextran and chitosan by layer-by-layer deposition. *Colloids Surf. A Physicochem. Eng. Asp.* **2014**, *450*, 121–129. [CrossRef]

7. Tassa, C.; Shaw, S.Y.; Weissleder, R. Dextran-coated iron oxide nanoparticles: A versatile platform for targeted molecular imaging, molecular diagnostics, and therapy. *Acc. Chem. Res.* **2011**, *44*, 842–852. [CrossRef]

8. Laurent, S.; Forge, D.; Port, M.; Roch, A.; Robic, C.; Elst, L.V.; Muller, R.N. Magnetic iron oxide nanoparticles: Synthesis, stabilization, vectorization, physico-chemical characterizations, and biological applications. *Chem. Rev.* **2008**, *108*, 2064–2110. [CrossRef]

9. Brunsen, A.; Utech, S.; Maskos, M.; Knoll, W.; Jonas, U. Magnetic composite thin films of Fe_xO_y nanoparticles and photocross linked dextran hydrogels. *J. Magn. Magn. Mater.* **2012**, *324*, 1488–1497. [CrossRef]

10. Hilger, I.; Hergt, R.; Kaiser, W.A. Use of magnetic nanoparticle heating in the treatment of breast cancer. *IEEE Proc. Nanobiotechnol.* **2005**, *152*, 33. [CrossRef]

11. Gilchrist, R.K.; Medal, R.; Shorey, W.D.; Hanselman, R.C.; Parrot, J.C.; Taylor, C.B. Selective inductive heating of lymph nodes. *Ann. Surg.* **1957**, *146*, 596–606. [CrossRef] [PubMed]

12. Babincova, M.; Babinec, F.; Bergemann, C. High-gradient magnetic capture of ferrofluids: Implications for drug targeting and tumor immobilization. *Z. Nat. C* **2001**, *56*, 909–911. [CrossRef] [PubMed]

13. Wang, Y.X.; Hussain, S.M.; Krestin, G.P. Superparamagneticiron oxide contrast agents: Physicochemical characteristics and applications in MR imaging. *Eur. Radiol.* **2001**, *11*, 2319–2331. [CrossRef] [PubMed]

14. Bonnemain, B. Superparamagneticagents in magneticresonanc e imaging: Physiochemical characteristics and clinical applications—A review. *J. Drug Target.* **1998**, *6*, 167–174. [CrossRef] [PubMed]

15. Prodan, A.M.; Iconaru, S.L.; Chifiriuc, M.C.; Bleotu, C.; Ciobanu, C.S.; Motelica-Heino, M.; Sizaret, S.; Predoi, D. Magnetic properties and biological activity evaluation of iron oxide nanoparticles. *J. Nanomater.* **2013**, *2013*, 1–7. [CrossRef]

16. Mahdavi, M.; Ahmad, M.B.; Haron, M.J.; Namvar, F.; Nadi, B.; Rahman, M.Z.A.; Amin, J. Synthesis, surface modification and characterisation of biocompatible magnetic iron oxide nanoparticles for biomedical applications. *Molecules* **2013**, *18*, 7533–7548. [CrossRef]

17. Xu, X.Q.; Shen, H.; Xu, J.R.; Xuc, J.; Li, X.J.; Xiong, X.M. Core-shell structure and magnetic properties of magnetite magnetic fluids stabilized with dextran. *Appl. Surf. Sci.* **2005**, *252*, 494–500. [CrossRef]

18. Syusaburo, H.; Masalatsu, H. Magnetic Iron Oxide–Dextran Complex and Process for Its Production. U.S. Patent No. 4,101,435, 18 July 1978.

19. Molday, R.S.; Mackenzie, D. Immunospecific ferromagnetic iron-dextran reagents for the labeling and magnetic separation of cells. *J. Immunol. Methods* **1982**, *52*, 353–367. [CrossRef]

20. Bunn, J.P.A.; Chan, D.C.; Kirpotin, D. Magnetic Microparticles. U.S. Patent No. 5,411,730, 2 May 1995.

21. Parka, J.Y.; Kima, J.S.; Nama, Y.S. Mussel-inspired modification of dextran for protein-resistant coatings of titanium oxide. *Carbohydr. Polym.* **2013**, *97*, 753–757. [CrossRef]

22. Shubayev, V.; Pisaniv, T.R.; Jin, S. Magnetic nanoparticles for theragnostics. *Adv. Drug Deliv. Rev.* **2009**, *61*, 467–477. [CrossRef]

23. Predoi, D.; Iconaru, S.L.; Predoi, M.V. Dextran-coated zinc-doped hydroxyapatite for biomedical applications. *Polymers* **2019**, *11*, 886. [CrossRef] [PubMed]

24. Teja, A.S.; Koh, P.Y. Synthesis, properties, and applications of magnetic iron oxide nanoparticles. *Prog. Cryst. Growth Charact. Mater.* **2009**, *55*, 22–45. [CrossRef]

25. Layek, S.; Pandey, A.; Pandey, A.; Verma, H.C. Synthesis of γ–Fe_2O_3 nanoparticles with crystallographic and magnetic texture. *Int. J. Eng. Sci. Technol.* **2010**, *2*, 33–39.

26. Hong, R.Y.; Feng, B.; Chen, L.L.; Liu, G.H.; Li, H.Z.; Zheng, Y.; Wei, D.G. Synthesis, characterization and MRI application of dextran-coated Fe_3O_4 magnetic nanoparticles. *Biochem. Eng. J.* **2008**, *42*, 290–300. [CrossRef]

27. Cullity, B.D. *Elements of X-ray Diffraction*, 3rd ed.; Prentice-Hall International: Upper Saddle River, NJ, USA, 2000.

28. Cumber, S.A.; Lead, J.R. Particle size distributions of silver nanoparticles at environmentally relevant conditions. *J. Chromatogr. A* **2009**, *1216*, 9099–9105. [CrossRef]

29. Huang, J.; Li, Q.; Sun, D.; Lu, Y.; Su, Y.; Yang, X.; Wang, H.; Wang, Y.; Shao, W.; He, N.; et al. Biosynthesis of silver and gold nanoparticles by novel sundried Cinnamomum camphora leaf. *Nanotechnology* **2007**, *18*. [CrossRef]

30. Khot, V.M.; Salunkhe, A.B.; Thorat, N.D.; Ningthoujam, R.S.; Pawar, S.H. Induction heating studies of dextran coated $MgFe_2O_4$ nanoparticles for magnetic hyperthermia. *Dalton Trans.* **2013**, *42*, 1249–1258. [CrossRef]

31. Kim, K.W. Biomedical applications of stereoscopy for three-dimensional surface reconstruction in scanning electron microscopes. *Appl. Microsc.* **2016**, *46*, 71–75. [CrossRef]

32. Ahmed, T.O.; Akusu, P.O.; Jonah, S.A.; Nasiru, R. Morphology and composition of nanocrystalline stabilized zirconia using SEM-EDS system. *Leonardo J. Sci.* **2011**, *19*, 81–92.

33. ImageJ. Available online: http://imagej.nih.gov/ij (accessed on 1 August 2019).

34. Rahbari, R.; Sheahan, T.; Modes, V.; Collier, P.; Macfarlane, C.; Badge, R.M. A novel L1 retrotransposon marker for HeLa cell line identification. *Biotechniques* **2009**, *46*, 277–284. [CrossRef]

35. Rezaei, M.; Mafakheri, H.; Khoshgard, K.; Montazerabadi, A.; Mohammadbeigi, A.; Oubari, F. The cytotoxicity of dextran-coated iron oxide nanoparticles on Hela and MCF-7 cancerous cell lines. *Iran. J. Toxicol.* **2017**, *11*, 31–36. [CrossRef]

36. Jeng, H.A.; Swanson, J. Toxicity of metal oxide nanoparticles in mammalian cells. *J. Environ. Sci. Health Part A* **2006**, *41*, 2699–2711. [CrossRef] [PubMed]

37. Kim, J.S.; Yoon, T.-J.; Yu, K.N.; Kim, B.G.; Park, S.J.; Kim, H.W.; Lee, K.H.; Park, S.B.; Lee, J.K.; Cho, M.H. Toxicity and tissue distribution of magnetic nanoparticles in mice. *Toxicol. Sci.* **2006**, *89*, 338–347. [CrossRef] [PubMed]
38. Karlsson, H.L.; Gustafsson, J.; Cronholm, P.; Möller, L. Size-dependent toxicity of metal oxide particles-a comparison between nano-and micrometer size. *Toxicol. Lett.* **2009**, *188*, 112–118. [CrossRef] [PubMed]
39. Dwivedi, S.; Siddiqui, M.A.; Farshori, N.N.; Ahamed, M.; Musarrat, J.; Al-Khedhairy, A.A. Synthesis, characterization and toxicological evaluation of iron oxide nanoparticles in human lung alveolar epithelial cells. *Colloids Surf. B* **2014**, *122*, 209–215. [CrossRef] [PubMed]
40. Karlsson, H.L.; Cronholm, P.; Gustafsson, J.; Moller, L. Copper oxide nanoparticles are highly toxic: A comparison between metal oxide nanoparticles and carbon nanotubes. *Chem. Res. Toxicol.* **2008**, *21*, 1726–1732. [CrossRef]
41. Villanueva, A.; Canete, M.; Roca, A.G.; Calero, M.; Veintemillas-Verdaguer, S.; Serna, C.J. The influence of surface functionalization on the enhanced internalization of magnetic nanoparticles in cancer cells. *Nanotechnology* **2009**, *20*, 115103. [CrossRef]
42. Ankamwar, B.; Lai, T.; Huang, J.; Liu, R.; Hsiao, M.; Chen, C.H. Biocompatibility of Fe_3O_4 nanoparticles evaluated by in vitro cytotoxicity assays using normal, glia and breast cancer cells. *Nanotechnology* **2010**, *21*, 075102. [CrossRef]

 *coatings*

Article

Starch/Poly (Glycerol-Adipate) Nanocomposite Film as Novel Biocompatible Materials

Domenico Sagnelli [1,*], Robert Cavanagh [2], Jinchuan Xu [1,3,4], Sadie M. E. Swainson [2], Andreas Blennow [3], John Duncan [5], Vincenzo Taresco [1] and Steve Howdle [1]

[1] School of Chemistry, University of Nottingham, University Park, Nottingham NG7 2RD, UK
[2] School of Pharmacy, University of Nottingham, University Park, Nottingham NG7 2RD, UK
[3] Department of Plant and Environment Sciences, University of Copenhagen, 1871 Frederiksberg C, Denmark
[4] School of Food Science and Engineering, South China University of Technology, Guangzhou 510640, China
[5] Lacerta Technology, Shepshed, Loughborough LE12 9GX, UK
* Correspondence: domenico.sagnelli@nottingham.ac.uk

Received: 11 July 2019; Accepted: 27 July 2019; Published: 30 July 2019

Abstract: Starch is one of the most abundant polysaccharides on the earth and it is the most important source of energy intake for humans. Thermoplastic starch (TPS) is also widely used for new bio-based materials. The blending of starch with other molecules may lead to new interesting biodegradable scaffolds to be exploited in food, medical, and pharmaceutical fields. In this work, we used native starch films as biopolymeric matrix carriers of chemo enzymatically-synthesized poly (glycerol-adipate) (PGA) nanoparticles (NPs) to produce a novel and biocompatible material. The prototype films had a crystallinity ranging from 4% to 7%. The intrinsic and thermo-mechanical properties of the composite showed that the incorporation of NPs in the starch films decreases the glass transition temperature. The utilization of these film prototypes as the basis for new biocompatible material showed promise, particularly because they have a very low or even zero cytotoxicity. Coumarin was used to monitor the distribution of the PGA NPs in the films and demonstrated a possible interaction between the two polymers. These novel hybrid nanocomposite films show great promise and could be used in the future as biodegradable and biocompatible platforms for the controlled release of amphiphilic and hydrophobic active ingredients.

Keywords: starch; nanoparticles; biomaterial; biocompatible; Caco-2; polyglycerol adipate; DMA; composites; polymer synthesis

---

1. Introduction

Starch is one of the most abundant polysaccharides on the earth and has a huge potential as a very well-defined raw material for functionalized biomaterials. Starch consists of two major polymers, amylose and amylopectin [1]. Amylopectin is a much larger molecule than amylose ($M_w = 1 \times 10^7$–1×10^9) and represents 65%–75% of starch. It has a highly branched structure, with α-(1–4)-linked D-glucose backbones and approximately 5% of α-(1–6)-linked branches [2]. Amylose is a quasi-linear polymer composed by α-(1–4)-linked glucose residues and has a molecular weight of approximately 1×10^5 to 1×10^6 Da [3]. Amylose has a small degree of α-(1–6)-linked branches. Amylopectin and amylose together form semi-crystalline and insoluble granules with an internal lamellar structure.

Thermoplastic starch (TPS) has been widely used as raw material for the production of novel biobased compostable materials [4–8]. Unfortunately, TPS forms a brittle material that is sensitive to water [8]. An option to improve the intrinsic characteristics of TPS is to blend or coat it with other biocompatible and bio-renewable polymers. Depending on the polymer used, the blend will have diverse functionalities and properties [4,9,10].

Bio-based poly (glycerol-adipate) (PGA) can be envisaged as a suitable candidate for starch co-polymeric-blends. In fact, PGA is a functionalizable, biocompatible, and biodegradable macromolecule [11–13]. Interestingly, PGA can be synthesized by a green poly-condensation reaction catalyzed by a lipase immobilized on acrylic beads (Novozyme 435) (Figure 1).

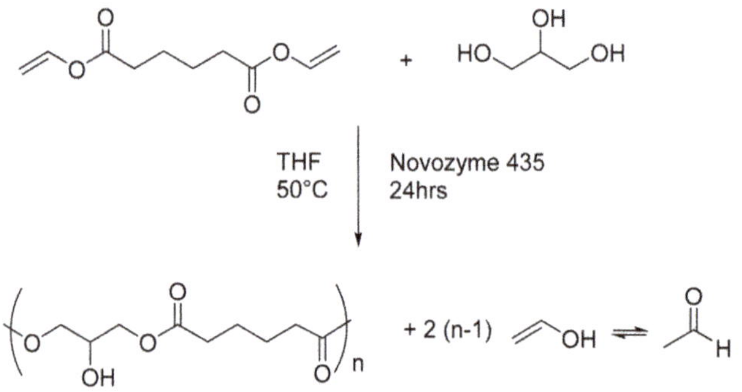

Figure 1. Polycondensation reaction of glycerol and dyvinil adipate catalyzed by Novozyme 435.

The enzymatic synthesis of poly-(glycerol adipate) (PGA) from divinyl adipate and glycerol is a simple and versatile strategy [11–13]. It is possible to exploit the chemo- and regio-selectivity of the enzyme to leave the secondary hydroxyl moiety unreacted, avoiding tedious and complicated protection/deprotection chemistry steps [12].

The reaction temperature is optimally performed at 50 °C to produce a highly linear polymer backbone motif with negligible branching [14–17]. Since the hydroxyl moieties are positioned in the outer shell of the nanoparticles (NPs), PGA will create low-force interactions with starch. This will avoid phase separations and allow the formation of a cohesive film. Furthermore, the utilization of a polysaccharide as the matrix for PGA nanoparticles will enable the formation of a possible biocompatible device for drug or flavor delivery.

In the present study, native starch extracted from barley was used as a base for cast biomaterial films. Because of the intrinsic brittleness of starch, the films were plasticized with glycerol. Furthermore, native starch films were used as a coating of PGA nanoparticles to produce a novel and biocompatible scaffold. The materials produced were characterized for their macro- and microstructure, chemical composition, crystallinity, thermos-mechanical properties, and biocompatibility.

2. Materials and Methods

2.1. Materials

Unless otherwise stated, all the chemicals used in this work were purchased from Sigma-Aldrich. Barley starch was kindly donated by Altia starch (Sweden). The human intestinal epithelial adenocarcinoma cell line, Caco-2, were obtained from the American Type Culture Collection (ATCC) and used between passages 45 and 50. Cells were cultured in Dulbecco's modified Eagle medium (DMEM) supplemented with 10% fetal bovine serum (FBS) at 37 °C in a humidified incubator with 5% CO_2. Cells were routinely grown in 75 cm^2 culture flasks to 70% confluence. Caco-2 cells were seeded at a density of 1×10^4 cells per well in 96-well plates for cytotoxicity experiments and cultured for 24 h before assaying.

2.2. Methods

2.2.1. Poly-(Glycerol Adipate)(PGA) Synthesis

Glycerol (125 mmol) and divinyl adipate (125 mmol) were poured into a 250 mL 3-neck round-bottomed flask and mixed with anhydrous 50 mL tetrahydrofuran (THF). The resulting system was heated in a tared oil bath. Lipase (1.1 g extracted ≥5000 U/g, recombinant, expressed in *Aspergillus niger*) was added into the mixture when the internal temperature of the reaction reached 40 to 45 °C and achieved a single homogeneous liquid phase. The stirring was set at 250 rpm using an overhead mechanical stirrer for 24 h. The reaction was stopped by filtration of the enzyme beads and THF evaporation. Any residual free enzyme in the dry fraction was deactivated by heating the product at 95 °C for 1 h. The final product was a pale yellow highly viscous liquid [12].

2.2.2. PGA N-Acyl-Tyrosine (Pgatyr) Coupling Reaction

A simple and well-established Steglich esterification was adopted to couple N-Acyl-Tyrosine to the free hydroxyl groups of PGA, aiming at 30% substitution of the total of the OH functionalities (Figure 2). Briefly, PGA (4.95 mmol) and DMAP (0.15 mmol) were added to anhydrous THF (20 mL) at room temperature in a round bottom flask under magnetic stirring until complete dissolution of the solids. Another THF solution (final volume 20 mL) was prepared by dissolving 1.20 mmol of DCC and 1.49 mmol of N-acyl Tyr. The two solutions were mixed and left to stir overnight under N_2 atmosphere. The resulting dicyclohexylurea insoluble precipitate was removed by centrifugation. The modified polymer was precipitated in a solution 0.1 M NaOH and twice in cold MeOH. The residual material was dried under reduced pressure to a stable weight.

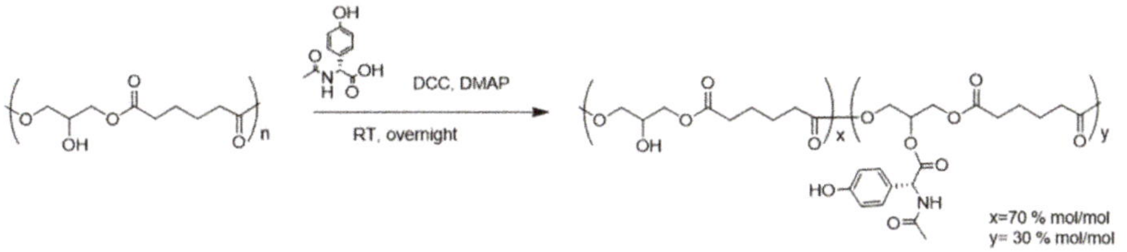

Figure 2. Coupling reaction scheme of Poly-(Glycerol Adipate)(PGA) and tyrosine.

2.2.3. Particle Size of PGA and PGATyr Nanoparticles (NPs)

PGA and PGATyr NPs were produced by the nanoprecipitation technique, reaching a final concentration of 1.5 mg/mL in the aqueous medium. Initially, the polymers were dissolved in acetone (2 mL) and added to ultrapure water (10 mL) under constant magnetic stirring (550 rpm) at Room Temperature (RT) (19 °C). The organic solvent was evaporated from the mixture while being agitated in a fume hood. Particle sizes were measured by dynamic light scattering (DLS, Zetasizer Nano ZS, Malvern Instruments, Malvern).

2.2.4. Enzymatic Degradation Assay

To evaluate the biodegradability of PGA and PGATyr, NPs (initial concentration 1.5 mg/mL) were diluted in PBS at a final concentration of 0.5 mg/mL. Lipase was prepared in PBS (pH 7.4) to a concentration of 10 mg/mL (215 units/mL). In total, 50 µL of the lipase mixture was added to the NP suspensions. DLS measurements were carried out for 180 sec after the enzyme addition at constant shaking at 37 °C to evaluate the swelling of the NPs in degradative conditions.

2.2.5. Melt Behavior Measurements

The melting behavior of the starch used in this study was measured using a high-pressure rheometer (Anton-Paar MCR series, Graz, Austria). In particular, a stirrer geometry was used and the rheometer program was as follows: 960 rpm for 90 s at 30 °C, followed by a pasting step at 170 rpm, a temperature ramp from 37 to 95 °C with a heating rate of 1 °C/min, an isotherm at 95 °C for 5 min, a final ramp from 95 to 50 °C at a cooling rate of 5 °C/min, for 5 min, and a final isotherm at 50 °C for 5 min.

2.2.6. Water Contact Angle (Θw)

Water contact angles (Θw) were measured at room temperature by employing a KSV Cam 200 (KSV Instruments Ltd, Helsinki, Finland) and angle pictures were analyzed with dedicated software (CAM200 1.1–KSV instruments). Four measurements were recorded for each polymer.

2.2.7. ATR FT-IR

ATR-IR spectra were acquired by means of an attenuated total reflection spectrophotometer (Agilent Technologies Cary 630 FTIR, Santa Clara CA, USA) equipped with a diamond single reflection ATR unit. Spectra were acquired with a resolution of 4 cm^{-1}, in the range 4000–650 cm^{-1} by acquiring 32 interferograms. Spectra were analyzed with the open access software SpectraGryph1.2.

2.2.8. Casting of Films

The native water content (%w) of starch (ST) was measured by thermogravimetric analysis and was taken into account for the final formulation. The films were casted with a final surface area of 10 mg/cm^2. The PGA nanoparticles were, respectively, 3% and 10% of starch (dry weight, d.w.). To achieve plasticization of the films, 25% of glycerol was included to the mixture.

Starch was suspended in de-ionized water. The solutions were mixed at 300 rpm with magnetic stirring following heating in an oil bath at 95 °C until gelatinization. After the gelatinization phase, the solutions were placed into an ice bath while being agitated. When a temperature of 50 °C was reached, glycerol and PGA NPs were added and mixed until full homogenization. The solutions were immediately casted in Teflon-coated petri dishes and the films dried at 50 °C in an oven with ventilation. Prior to any analysis, the film prototypes were equilibrated to stabilize the moisture content in a desiccator containing a saturated solution of potassium chloride (relative humidity 85% ± 1%).

2.2.9. X-Ray Scattering

The films were loaded in a sealed holder to minimize evaporation during the measurement. X-ray measurements were conducted using a Panalytical Xpert Pro (Nottingham, UK).

The radially averaged intensity is given as a function of the scattering vector, $q = 4\pi \sin \theta/\lambda$, where λ is the wavelength and 2θ is the scattering angle. The samples were measured in the WAXS setting (wide-angle X-ray scattering) to an upper 2θ value of 35° or 2.86 Å. The exposure time was 400 s/step with a step size of (0.0131303°) [5].

2.2.10. Dynamic Mechanical Analysis

Dynamic mechanical analysis (DMA) with a temperature gradient was performed in tension mode with a displacement of 0.005 mm and frequencies of 1 and 10 Hz. A standard heating rate of 3 °C min^{-1} was used and a ramp from −50 to 120 °C. The experiments were performed on prototypes with a length of 10 mm [7]. The glass transition temperature was estimated by comparing the derivative function of the storage modulus and the tan delta (tanδ) peak.

For measurements in a humidity-controlled environment, the temperature was kept constant at 37 °C and the relative humidity (RH) was increased up to 80% at a rate of 0.3 %RH/min and kept at 80% until equilibration.

2.2.11. Confocal Microscopy

Confocal laser scanning microscopy (CLSM) was performed using a Leica system (Leica SP5-X, Leica Microsystems) equipped with x63 water immersion objectives. Films were prepared as previously described. PGA and PGA-Tyr NPs were prepared by nanoprecipitation to a final concentration of 1.5 mg/mL. Coumarin 30, used as fluorophore, was dissolved in methanol to a concentration of 1 mg/mL and the polymer was dissolved in acetone at the same concentration. Coumarin 30 solution (15 µL per mg of polymer) was added to the polymer solution and the resultant solution added dropwise to ultrapure water under magnetic stirring. The nanoparticle suspension was immediately transferred to a dialysis membrane with a 3.5 kDa MWCO (Mw cut off, Regenerated Cellulose Dialysis Tubing, Type T1, Fisherbrand) and dialyzed against water for 4 h to remove the acetone, methanol, and any free coumarin 30. The acetone was evaporated, and particles were dialyzed against water for 3 h. Images were analyzed with the LAS X 3.0 software.

2.2.12. Cytotoxicity Test

Assessment of cytotoxicity was performed in accordance with ISO standard 10993-5 via an extract test to evaluate the biocompatibility of any leachable material by-products by the simulation of clinical application [18]. The PrestoBlue cell viability assay (ThermoFisher) was used to probe mitochondrial function and the detection of lactose dehydrogenase (LDH) release was used to measure cell membrane integrity (Sigma-Aldrich, TOX7 kit, St. Louis MO, USA). Films were UV sterilized for 20 min and placed in DMEM (Dulbecco minimal essential medium) with 10% fetal bovine serum (FBS) (3 cm^2 per ml) for 24, 48 and 72 h and incubated at 37 °C in a humidified incubator with 5% CO_2. The resulting solutions were applied to cells (100 µL per well) and incubated for 48 h. A negative control with DMEM and a positive control (0.1% Triton X-100) were also included in the cytotoxicity experiments. Following exposure, 50 µL of supernatant were collected per well for analysis of LDH content. Cells were washed twice with phosphate-buffered saline (PBS) and 100 µL 10% (v/v) PrestoBlue reagent diluted in DMEM was applied per well for 60 min. The resulting fluorescence was measured at 560/600 nm (λex/λem). Relative metabolic activity was calculated by setting values from the negative control as 100% and positive control values as 0% metabolic activity. Assessment of LDH release was performed according to the manufacturer's instructions and involved adding 100 µL of LDH reagent to the collected supernatant samples and incubating them at room temperature shielded from light for 25 minutes. Absorbance was measured at 492 nm. Relative LDH release was calculated with the negative control absorbance at 492 nm taken as 0%, and the positive control, assumed to cause total cell lysis, as 100%.

3. Results and Discussion

3.1. Native Starch Melt and Crystallinity Characterization

Before the production of the films, the native starch was analyzed for its melting behavior in water to allow a precise design of the melting process. In particular, the 10% suspension of starch granules was processed using a high-pressure sealed rheometer to avoid any evaporation of water (Figure 3). This rheometer allowed a better understanding of the starch behavior during melting as compared to standard rapid visco analysers, which have open vessels, thus risking significant water evaporation.

The measurement showed a pasting viscosity of 18.8 cP at a temperature of 88 °C. The hot peak viscosity was measured at 95 °C. During the isotherm, a breakdown of viscosity was detected at 95 °C, indicating complete melting of the starch. During the fast cooling phase, when a temperature of 50 °C was reached, the melt started to form a gel, with a peak viscosity of 1300 cP. These measurements allowed the design of a melting profile for the production of the films [6].

Since starch has a semi-crystalline nature, it was first characterized using wide angle X-ray scattering (WAXS) to identify its diffraction pattern following full hydration [5]. The initial crystallinity also provided data to calculate the degree of disruption of the native crystallinity in the films. The results showed that the native starch possessed an average degree of crystallinity (20%) and was found to be a common mixture of A-type and B-type polymorphs, with a V-type peak centered at 20 (2θ) [19]. These data prove that the starch used, even if it was industrially extracted, did not undergo any transformation of the crystalline and amorphous structure.

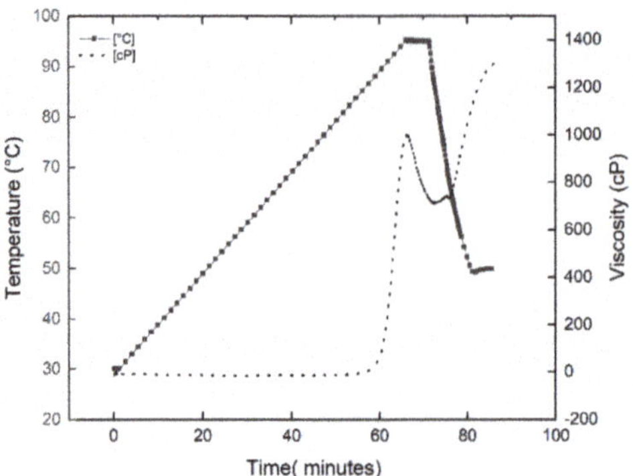

Figure 3. Melting profile measurement of 10% starch suspension using a high-pressure rheometer.

3.2. Synthesis and Characterization of Nanoparticles

PGA and PGATyr NPs were formulated via the simple nanoprecipitation technique, using acetone as the organic phase. Due to the intrinsic amphiphilic structure established among the backbone and the side groups of these polymers, the poly-(glycerol adipate) can favorably self-assemble in water without the use of surfactants, thus reducing undesired side effects. Before the preparation of the composite films, an enzymatic dynamic light scattering (DLS) screening assay to confirm the biodegradability of PGA and PGATyr NPs was performed [20]. PGATyr NPs (145 nm) were smaller than the NPs produced from bare PGA (175 nm), confirming the existence of better packing in the hydrophobic core, supposedly due to the π–π interactions among the aromatic moieties [13].

NPs' size alterations were recorded in the presence of lipase against NPs enzyme-free controls. It has been suggested that the increase in particle size represented by the swelling of the NPs is a sign of degradation when interacting with enzymes [21]. Remarkable swelling in terms of hydrodynamic size was observed for both materials. However, a different swelling pattern was observed for the two polyesters after 3 h of contact with lipase (Figure 4). In fact, PGATyr reached half of the size of the pure PGA NPs. These differences may be related to a lower enzyme accessibility because of steric hindrance caused by different polymer chains spatial arrangement or chains packing and interactions linkable to the π–π stacking interactions amongst the aromatic tyrosyl residues [13].

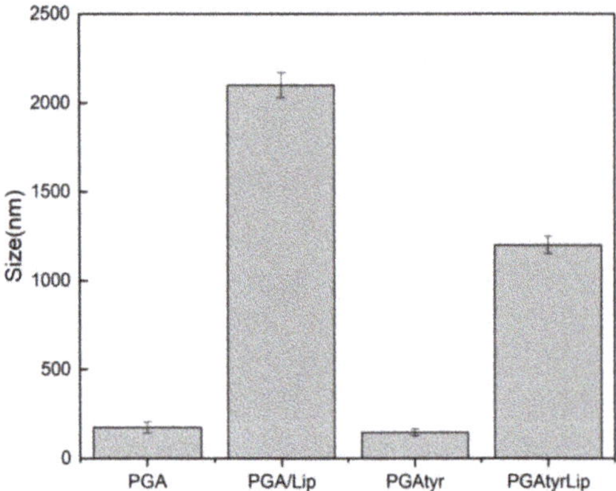

Figure 4. Size distribution and lipase assay (after 3 h of contact with the lipase).

3.3. Crystallinity, Contact Angle, and Microstructures of Films

After the melting and crystallinity characterization of native starches, films were prepared by casting. Before any analysis, the films were equilibrated in an environment with controlled relative humidity (RH: 85%). The crystallinity showed a decrease for all the films as compared to native starch (Table 1). In particular, the different crystalline polymorphs and their relative contribution to the overall crystallinity were altered following recrystallization during the drying process (Table 1). The crystallinity of the films was severely reduced by the heat treatment. In fact, the crystallinity was three-fold lower for the films with unmodified PGA NPs and four-fold lower when the NPs were grafted with tyrosine. Furthermore, for all films, no A-type polymorphism was individuated, confirming the complete melt of the crystalline structure during the heating treatment [5]. The crystalline polymorphs detected in the films are due to the natural process of recrystallization of the starch. Interestingly, the addition of NPCs in the starch films allowed the growth of a similar relative distribution of Vh- and B-type polymorphs.

Table 1. Percentage of crystallinity and the contact angle of native starch and cast films. Vh, B, and A denotes the crystalline polymorphs.

Sample	Cryst. (%)	Vh (%)	B (%)	A (%)	Contact Angle (°)
StNative	20	5	18	77	
StOn	6	37	63	0	102 ± 1
StGl	6	20	80	0	88 ± 2
StPGA3	7	52	48	0	nd
StPGA10	7	47	53	0	57 ± 0.5
StPGAtyr3	5	57	39	0	nd
StPGAtyr10	4	50	50	0	32 ± 1

The stacked IR spectra of the four starch films (Figure 5) demonstrate that the introduction in the starch matrix of both glycerol and the polymeric NPs led to variations in the hydrogen bond network depicted by the variation in the wave number of the OH stretching in the region between 3600 and 3200 cm^{-1}. In the spectra of starch mixed with pure PGA (StPGA) and starch mixed with the PGA modified with tyrosine (StPGATyr), the appearance of a peak in the region around 1700 cm^{-1} is noticeable. This band is related to the stretching of the ester group, confirming the blending of the NPs within the film. Minor peak shifting and splitting can be observed in the area around 3000 to 2800 cm^{-1} and 1050 to 950 cm^{-1} related to the CH stretching and the C–O–C stretching, respectively. Due to the tight hydrogen bond network among the hydroxyl groups in the starch molecules, an extremely smooth and hydrophobic surface was observed (Figure 6 and Table 1). This phenomenon is likely due to the masking of the polar OH, involved in the intra- and intermolecular H-bonding, within the starch matrix. However, by adding glycerol, PGA, or PGATyr NPs, a change in the recrystallization processes was expected and hence a perturbation of this H-bonding network. The addition of these molecules resulted in a drop in the value of the water contact angle (Table 1). The variation in the water contact angle was more prominent after the addition of the two polymeric NPs compared to the variation related to the presence of glycerol (Table 1). Interestingly, a direct comparison of the water contact angle and smoothness was observed (Figure 6 and Table 1). The films with only starch (StOn) and starch mixed with glycerol (StGl) showed a smooth surface and consequently high contact angle values of 102° and 88°, respectively. On the other hand, the addition of the polymeric NPs tremendously affected the integrity of the starch surface film (Figure 6). A remarkable drop in the water contact angle (57° and 32°) can be ascribed to these superficial alterations. It can be speculated that both the aliphatic and aromatic hydroxyl groups in the starch films containing PGAtyr NPs (StPGATyr) may deeply alter the nature and arrangement of the H-bonding network within the film. The films containing StPGATyr typically showed a lower relative crystallinity.

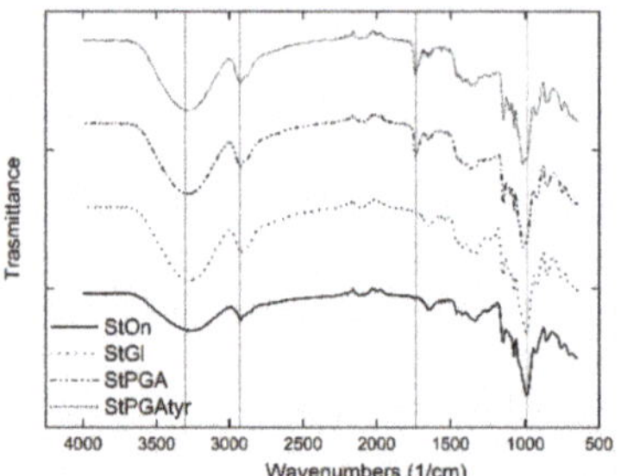

Figure 5. Stacked IR spectra of StOn, StGl, StPGA, and StPGATyr in the region 4000–650 cm^{-1}.

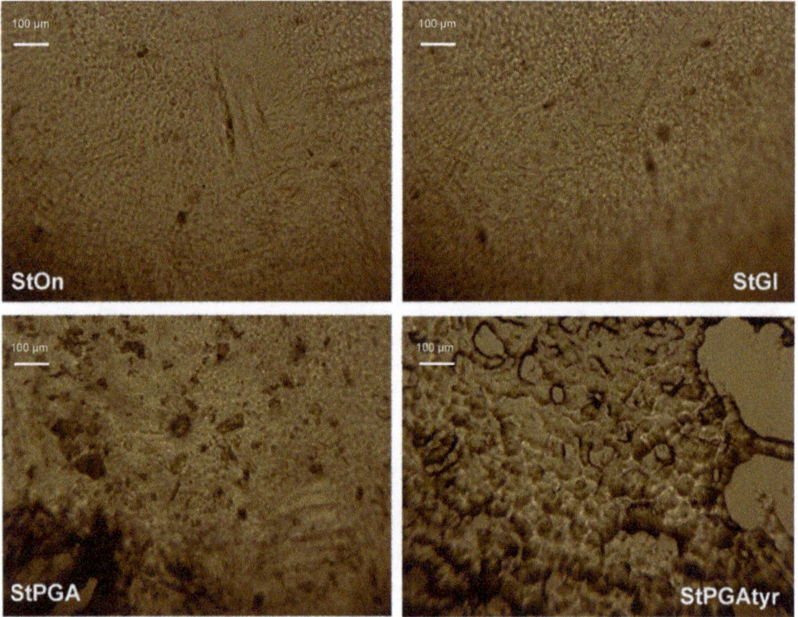

Figure 6. Optical microscopy pictures showing different surface smoothness of StOn, StGi, StPGA, and StPGATyr films.

To monitor the distribution of the poly(glycerol adipate) in the films, the NPs were loaded with a fluorescent dye (coumarin), which was dispersed into the films (Figure 7). Two phases were easily visualized. The films containing the 3% NPs showed a fibril-like structure formed between starch (dark phase) and PGA (blue phase) (Figure 7B). This demonstrates that PGA NPs were stable during the drying of the films and the coumarin did not leak during fabrication and testing. Furthermore, the interaction between starch and PGA NPs was demonstrated. When the concentration of NPs was increased to 10%, the fibril structure was lost, probably due to the aggregation of the NPs (Figure 7A).

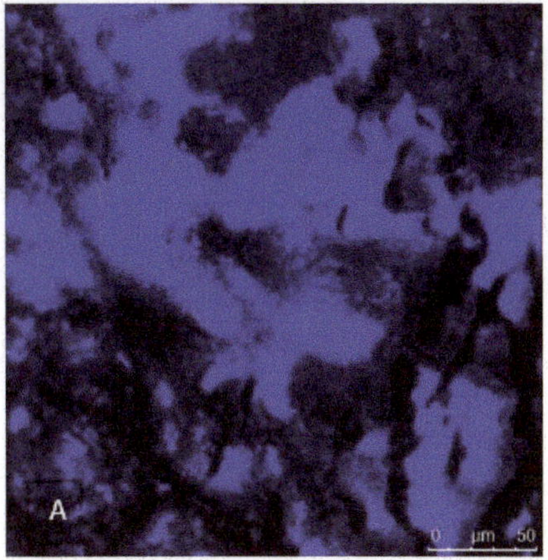
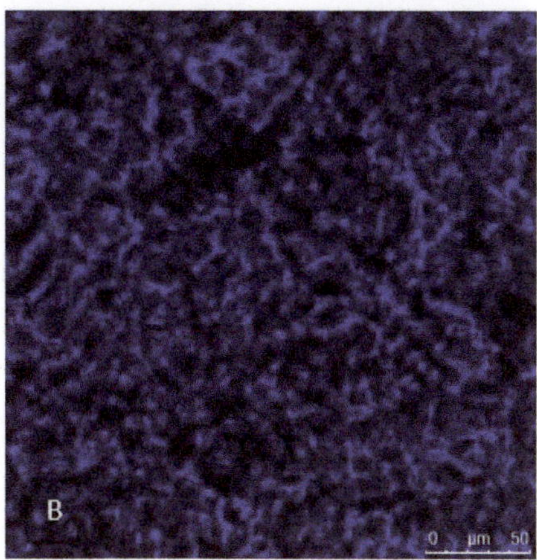

Figure 7. Confocal microscopy pictures showing the NPs distribution in the starch films. (**A**) StPGATyr 10%; (**B**) StPGATyr 3%.

3.4. Thermo-Mechanical Properties

Dynamic mechanical analysis was performed using controlled temperature and humidity scans. Firstly, the films were scanned to define the Tg of the blends equilibrated at 80% RH. In particular, the addition of NPCs showed a bimodal DMA trace distribution, with zones richer in NPs that vibrated at lower temperaturea. The starch-rich pockets showed a higher Tg as compared to the starch alone (Table 2). The glass transition temperature of free PGA could not be recorded in these experiments. This was caused by the fact that PGA assembled in NPs embedded in a complex three-dimensional and semi-crystalline matrix. Hence, the glass transition of the poly-ester changed under these conditions, increasing to about 37 °C.

Table 2. Glass transition temperature measured using a dynamic mechanical temperature analyser and dynamic mechanical humidity analyser.

Film	Tg	Tg-RH
starch	66	37 °C – 80%
StPGA3	38/-	
StPGA3tyr	-/80	37 °C – 57%
StPGA10	37/80	
StPGA10tyr	37/80	

StGl and StPGATyr were further analyzed in a humidity-controlled environment (Figure 8). In particular, these results showed that the PGA NPs changed their sensitivity to the water of the starch films (Table 2). The glass transition of starch at 37 °C was detected only when the relative humidity reached 80%. On the other hand, StPGAtyr3 films showed a Tg transition at a relative humidity of 57% relative humidity, indicating a tendency in lowering the Tg in a lower-moisture environment.

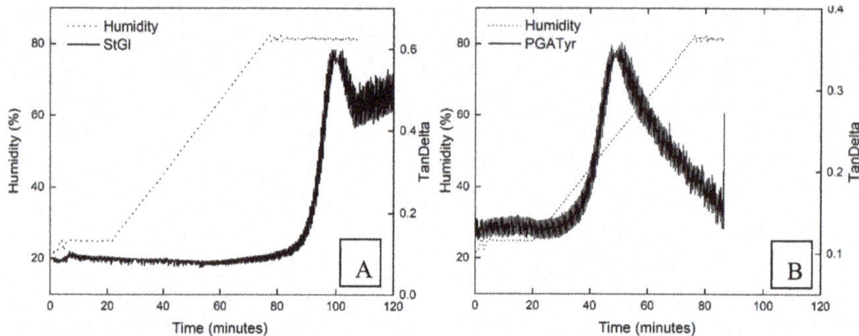

Figure 8. Dynamic mechanical analysis with humidity control of (**A**) starch film with 25% glycerol and (**B**) starch film with 25% glycerol and 3% PGA NPs.

3.5. Biocompatibility Tests

Because starch is a biopolymer digested in the human intestine, biocompatibility tests were performed on Caco-2 cells, thus giving insight into the possible utilization of these films as a biocompatible material for drug or bio-actives release.

Cytotoxicity analysis in Caco-2 cells of film extracts generated from different incubation times with culture medium demonstrated that the extracts from the materials can be considered biocompatible, as no loss in cellular metabolic activity or damage to cell membranes was experienced following incubation with cells for 48 h (Figure 9). Indeed, compared to the negative control (DMEM), increases in metabolic activity (approximately 10% to 20%) were observed in various groups, including the starch and StGl and StOn films (Figure 9); an effect likely related to the presence of free starch in the film extracts and the subsequent availability of this carbohydrate for cellular metabolism.

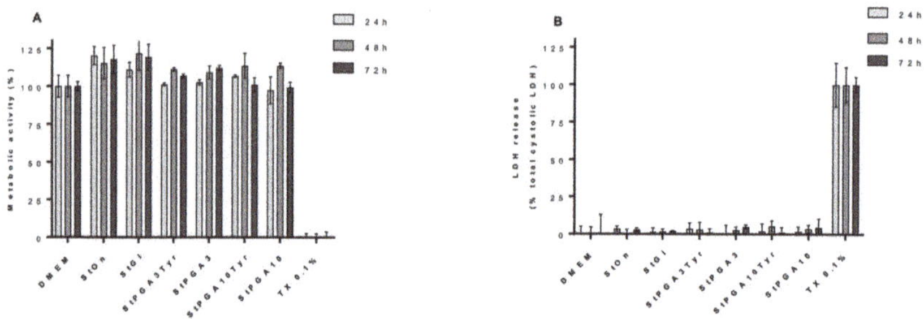

Figure 9. Cytocompatibility of PGA films assessed via (**A**) cellular metabolic activity and (**B**) LDH release in Caco-2 intestinal epithelial cells. Films were incubated with DMEM (3 cm^2 per mL) for 24, 48, or 72 h and the resulting solutions were applied to cells for 48 h. Data are presented as mean ± S.D (n = 3).

These data demonstrate that starch-based films can be used as a base for biomaterials and for drug delivery. In fact, the films containing NPs did increase the metabolic activity.

4. Conclusions

Casted barley starch films were tested as bio polymeric carriers for poly-(glycerol adipate)-based nanoparticles. The native starch granules showed typical behavior during characterization of the melting profile and crystallinity. When the films were prepared, the overall crystallinity varied. In fact, for all the films, we identified a decrease in crystallinity down to a minimum of 4%. Furthermore, the films showed a mixture of Vh- and B-type polymorphs. Interestingly, the films with

NPs showed a ratio between B and Vh of approximately 1:1. This result may be an indication of the formation of starch/PGA complexes.

The integration of nanoparticles in glycerol-plasticized films showed an interesting effect on the glass transition temperature. In fact, it was possible to measure variation in Tg and an increase in moisture sensitivity as well as in the final wettability of the film surface and superficial roughness.

In conclusion, the films showed a good capability for stimulating the growth of Caco-2 cells. Our work proves that starch has the potential to be a good biocompatible material and a coating polymer for nanoparticles. Furthermore, the films created are a possible alternative as a coating for food materials and the delivery of bioactive molecules (e.g., flavors).

Author Contributions: Conceptualization, D.S., V.T.; data curation, D.S., R.C., J.D., and V.T.; formal analysis, D.S., R.C., J.D., and V.T.; funding acquisition, D.S.; investigation, D.S., R.C., J.X., S.M.E.S., J.D., and V.T.; supervision, S.M.H.; writing—original draft, D.S. and V.T.; writing—review and editing, D.S., A.B., V.T., and A.M.H.

Funding: This research was funded by Danish council for independent research, grant number 7026-00060B.

Acknowledgments: The authors thank the Danish Council for Independent Research Technology, for the invaluable help guiding me towards this work. The authors also thank the Center for Advanced Bioimaging, Faculty of Science, the University of Copenhagen for use of their microscopy facilities.

Conflicts of Interest: The authors declare no conflict of interest. The funders had no role in the design of the study; in the collection, analyses, or interpretation of data; in the writing of the manuscript, or in the decision to publish the results.

References

1. Pérez, S.; Bertoft, E. The molecular structures of starch components and their contribution to the architecture of starch granules: A comprehensive review. *Starch/Staerke* **2010**, *62*, 389–420. [CrossRef]
2. Blennow, A.; Engelsen, S.B.; Nielsen, T.H.; Baunsgaard, L.; Mikkelsen, R. Starch phosphorylation: A new front line in starch research. *Trends Plant Sci.* **2002**, *7*, 445–450. [CrossRef]
3. Blennow, A.; Mette Bay-Smidt, A.; Bauer, R. Amylopectin aggregation as a function of starch phosphate content studied by size exclusion chromatography and on-line refractive index and light scattering. *Int. J. Biol. Macromol.* **2001**, *28*, 409–420. [CrossRef]
4. Sagnelli, D.; Kirkensgaard, J.J.K.; Giosafatto, C.V.L.; Ogrodowicz, N.; Kruczał, K.; Mikkelsen, M.S.; Maigret, J.E.; Lourdin, D.; Mortensen, K.; Blennow, A. All-natural bio-plastics using starch-betaglucan composites. *Carbohydr. Polym.* **2017**, *172*, 237–245. [CrossRef] [PubMed]
5. Sagnelli, D.; Hebelstrup, K.H.; Leroy, E.; Rolland-Sabaté, A.; Guilois, S.; Kirkensgaard, J.J.K.; Mortensen, K.; Lourdin, D.; Blennow, A. Plant-crafted starches for bioplastics production. *Carbohydr. Polym.* **2016**, *152*, 398–408. [CrossRef] [PubMed]
6. Sagnelli, D.; Hooshmand, K.; Kemmer, G.C.; Kirkensgaard, J.J.K.; Mortensen, K.; Giosafatto, C.V.L.; Holse, M.; Hebelstrup, K.H.; Bao, J.; Stelte, W.; et al. Cross-Linked Amylose Bio-Plastic: A Transgenic-Based Compostable. *Int. J. Mol. Sci.* **2017**, *18*, 2075.
7. Lourdin, D.; Coignard, L.; Bizot, H.; Colonna, P. Influence of equilibrium relative humidity and plasticizer concentration on the water content and glass transition of starch materials. *Polymer (Guildf)* **1997**, *38*, 5401–5406. [CrossRef]
8. Follain, N.; Joly, C.; Dole, P.; Bliard, C. Mechanical properties of starch-based materials. I. Short review and complementary experimental analysis. *J. Appl. Polym. Sci.* **2005**, *97*, 1783–1794. [CrossRef]
9. Montero, B.; Rico, M.; Rodríguez-Llamazares, S.; Barral, L.; Bouza, R. Effect of nanocellulose as a filler on biodegradable thermoplastic starch films from tuber, cereal legume. *Carbohydr. Polym.* **2017**, *157*, 1094–1104. [CrossRef] [PubMed]
10. Jantanasakulwong, K.; Leksawasdi, N.; Seesuriyachan, P.; Wongsuriyasak, S.; Techapun, C.; Ougizawa, T. Reactive blending of thermoplastic starch, epoxidized natural rubber and chitosan. *Eur. Polym. J.* **2016**, *84*, 292–299. [CrossRef]
11. Kallinteri, P.; Higgins, S.; Hutcheon, G.A.; St. Pourçain, C.B.; Garnett, M.C. Novel functionalized biodegradable polymers for nanoparticle drug delivery systems. *Biomacromolecules* **2005**, *6*, 1885–1894. [CrossRef] [PubMed]

12. Taresco, V.; Creasey, R.G.; Kennon, J.; Mantovani, G.; Alexander, C.; Burley, J.C.; Garnett, M.C. Variation in structure and properties of poly(glycerol adipate) via control of chain branching during enzymatic synthesis. *Polymer (Guildf)* **2016**, *89*, 41–49. [CrossRef]

13. Taresco, V.; Suksiriworapong, J.; Styliari, I.D.; Argent, R.H.; Swainson, S.E.; Booth, J.; Turpin, E.; Laughton, C.A.; Burley, J.C.; Alexander, C.; et al. New N-acyl amino acid-functionalized biodegradable polyesters for pharmaceutical and biomedical applications. *RSC Adv.* **2016**, *6*, 109401–109405. [CrossRef]

14. Bilal, M.H.; Hussain, H.; Prehm, M.; Baumeister, U.; Meister, A.; Hause, G.; Busse, K.; Mäder, K.; Kressler, J. Synthesis of poly(glycerol adipate)-g-oleate and its ternary phase diagram with glycerol monooleate and water. *Eur. Polym. J.* **2017**, *91*, 162–175. [CrossRef]

15. Wersig, T.; Krombholz, R.; Janich, C.; Meister, A.; Kressler, J.; Mäder, K. Indomethacin functionalised poly(glycerol adipate) nanospheres as promising candidates for modified drug release. *Eur. J. Pharm. Sci.* **2018**, *123*, 350–361. [CrossRef] [PubMed]

16. Taresco, V.; Suksiriworapong, J.; Creasey, R.; Burley, J.C.; Mantovani, G.; Alexander, C.; Treacher, K.; Booth, J.; Garnett, M.C. Properties of acyl modified poly(glycerol-adipate) comb-like polymers and their self-assembly into nanoparticles. *J. Polym. Sci. Part A Polym. Chem.* **2016**, *54*, 3267–3278. [CrossRef] [PubMed]

17. Suksiriworapong, J.; Taresco, V.; Ivanov, D.P.; Styliari, I.D.; Sakchaisri, K.; Junyaprasert, V.B.; Garnett, M.C. Synthesis and properties of a biodegradable polymer-drug conjugate: Methotrexate-poly(glycerol adipate). *Colloids Surf. B Biointerfaces* **2018**, *167*, 115–125. [CrossRef]

18. Wang, M.O.; Etheridge, J.M.; Thompson, J.A.; Vorwald, C.E.; Dean, D.; Fisher, J.P. Evaluation of the in vitro cytotoxicity of cross-linked biomaterials. *Biomacromolecules* **2013**, *14*, 1321–1329. [CrossRef]

19. Vasanthan, T.; Hoover, R. Barley Starch: Production, Properties, Modification and Uses. In *Starch*; Elsevier: Amsterdam, The Netherlands, 2009; pp. 601–628.

20. Marin, E.; Briceño, M.I.; Caballero-George, C. Critical evaluation of biodegradable polymers used in nanodrugs. *Int. J. Nanomed.* **2013**, *8*, 3071–3091. [PubMed]

21. Colombo, C.; Dragoni, L.; Gatti, S.; Pesce, R.M.; Rooney, T.R.; Mavroudakis, E.; Ferrari, R.; Moscatelli, D. Tunable degradation behavior of PEGylated polyester-based nanoparticles obtained through emulsion free radical polymerization. *Ind. Eng. Chem. Res.* **2014**, *53*, 9128–9135. [CrossRef]

 *coatings* 

Article

S-Layer Protein Coated Carbon Nanotubes

Andreas Breitwieser [1], Philipp Siedlaczek [2], Helga Lichtenegger [2], Uwe B. Sleytr [1]
and Dietmar Pum [1,*]

1   Department of Nanobiotechnology, Institute of Biophysics, University of Natural Resources and Life
    Sciences Vienna, Muthgasse 11, 1190 Vienna, Austria
2   Department of Material Sciences and Process Engineering, Institute of Physics and Materials Science,
    University of Natural Resources and Life Sciences Vienna, Peter-Jordan-Straße 82, 1190 Vienna, Austria
*   Correspondence: dietmar.pum@boku.ac.at; Tel.: +43-47654-80360

Received: 16 July 2019; Accepted: 31 July 2019; Published: 2 August 2019

Abstract: Carbon nanotubes (CNTs) have already been considered for medical applications due to their small diameter and ability to penetrate cells and tissues. However, since CNTs are chemically inert and non-dispersible in water, they have to be chemically functionalized or coated with biomolecules to carry payloads or interact with the environment. Proteins, although often only randomly bound to the CNT surface, are preferred because they provide a better biocompatibility and present functional groups for binding additional molecules. A new approach to functionalize CNTs with a closed and precisely ordered protein layer is offered by bacterial surface layer (S-layer) proteins, which have already attracted much attention in the functionalization of surfaces. We could demonstrate that bacterial S-layer proteins (SbpA of *Lysinibacillus sphaericus* CCM 2177 and the recombinant fusion protein $rSbpA_{31\text{-}1068}GG$ comprising the S-layer protein and two copies of the IgG binding region of Protein G) can be used to disperse and functionalize oxidized multi walled CNTs. Following a simple protocol, a complete surface coverage with a long-range crystalline S-layer lattice can be obtained. When $rSbpA_{31\text{-}1068}GG$ was used for coating, the introduced functionality could be confirmed by binding gold labeled antibodies via the IgG binding domain of the fusion protein. Since a great variety of functional S-layer fusion proteins has already been described, our new technology has the potential for a broad spectrum of functionalized CNTs.

Keywords: S-layer protein; carbon nanotubes; functionalization; non-covalent; IgG binding domain; dispersion; aqueous solution

---

1. Introduction

Since their discovery, carbon nanotubes (CNTs) have already been intensively investigated and characterized in material sciences due to their outstanding mechanical, electrical, and thermal properties [1–4]. While some of the developments of new applications are still in progress, others have already been materialized into new products. Moreover, CNTs have also been considered for several medical applications due to their small diameter and ability to penetrate cells and tissues [5]. However, since pristine CNTs are chemically inert and not dispersible in water or organic solvents, they have to be functionalized or modified to carry payloads or interact with the environment [6–11].

Proteins bound to the surface of CNTs are preferred in life-sciences because they provide a better biocompatibility and offer functional groups that may either be used for binding additional molecules in biosensor applications [12–19] or enable further chemical modifications, e.g., for the delivery of drugs, DNA and genes [20,21]. Nevertheless, although several proteins, such as bovine serum albumin (BSA), have already been successfully attached to CNTs through various physical or chemical methods, high resolution microscopical studies have demonstrated that their arrangement and density on the CNT surface and consequently the availability of functional groups varies considerably [5].

An alternative and better controlled approach to functionalize CNTs with an additionally closed and precisely ordered protein layer is offered by bacterial surface layer (S-layer) proteins which have already attracted much attention in the functionalization of surfaces as well as supporting structures for biomembranes [22–25].

S-layer proteins are one of the most abundant biopolymers on earth and form the outermost cell envelope component in a broad range of bacteria and archaea (Figure 1a) [22]. In addition to the surface of bacterial cells, S-layer proteins have the natural capability to reassemble into crystalline monomolecular arrays on solid supports, at the air-water interface, planar lipid films, liposomes, emulsomes, nanocapsules, and nanoparticles (Figure 1b) [23].

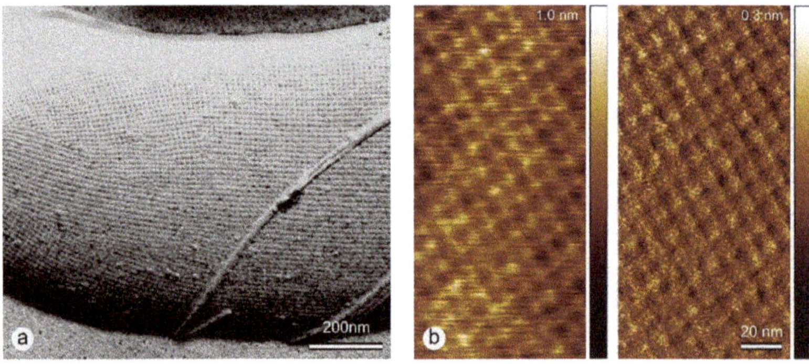

Figure 1. (**a**) TEM micrograph of a freeze-etched and metal shadowed preparation of a bacterial cell of *Lysinibacillus sphaericus* CCM 2177 with an S-layer as the outermost cell envelope component. The numerous lattice faults are a consequence of the bending of the S-layer lattice at the rounded cell poles. In addition, the rope-like structures are the flagella of the bacterial cell. (Reproduced from Reference [26] with permission from the Royal Society of Chemistry.) (**b**) AFM images of a monolayer of the SbpA S-layer lattice of *Lysinibacillus sphaericus* CCM 2177 on a silicon wafer. (Height image (left) and deflection error image (right)). The S-layer lattice shows square lattice symmetry. Unit cell size is 13.1 × 13.1.

S-layers are isoporous protein mesh works with unit cell sizes in the range of 3 to 30 nm, thicknesses of 5 to 10 nm (up to 70 nm in archaea), and pore sizes of 2 to 8 nm (Figure 1b). Since S-layers are composed of a single protein or glyco-protein species they may be considered as the simplest biological membranes developed in the course of evolution.

In particular, the formation of monolayers on technologically important substrates, such as silicon or glass, was always a major concern for the development of affinity matrices, biosensing layers or the development of organic-inorganic hybrid architectures [24,27]. In this context, it was seen as a further challenge to investigate the reassembly of S-layer proteins on CNTs and learn from nature how these new hybrid architectures may be used to develop a possible next generation of biological sensing layers. Key to such developments are S-layer (fusion) proteins [22,28] that, on the one hand, have retained the natural self-assembly properties of the wild-type proteins and, on the other hand, are endowed with particularly tailored bio-reactive domains that allow a highly specific and sensitive functionalization of surfaces. As a matter of fact, functional groups on the protein lattice are arranged in well-defined positions and orientations [22,24]. Examples are affinity matrices with S-layer fusion proteins carrying the immunoglobulin G (IgG) binding domains of Protein A or Protein G [29–31] or the green fluorescent protein (GFP and its variants) for Förster- or Fluorescence-energy transfer (FRET) pairs in DNA-hairpin sensors [32]. In this context, it has to be stressed that the high binding capacity of S-layer proteins would also be retained after intra- and intermolecular crosslinking (e.g., by glutardialdehyde or Dimethyl-pimelimidatedihydrochloride (DMP)). It has been shown that cross-linking enhances the mechanical and chemical stability of S-layers (e.g., at sudden pH changes or higher temperatures) considerably [24].

To our knowledge, this work describes for the first time the reassembly of an S-layer protein; in particular, of SbpA, the S-layer protein from *Lysinibacillus sphaericus* CCM 2177 [33,34] (identical to *Lysinibacillus sphaercius* ATCC 4525, see Reference [35]) with its characteristic square (p4) lattice symmetry on CNTs. With respect to the unit cell size of SbpA with 13.1 × 13.1 nm, we decided to work with multiwalled nanotubes (MWNTs) with diameters ranging from 50–90 nm since the diameters of single- and double-walled nanotubes (typically below 10 nm) might be too small (Figure 2). The addition of S-layer protein to aggregated CNTs led to an instantaneous dispersion of the CNTs. According to the literature, amphiphilic molecules, such as S-layer proteins, are suitable to disperse CNTs in water by shielding their highly hydrophobic surface [8,36]. Moreover, we assume that this effect might be emphasized by the fact that S-layer recrystallization follows a two-step non-classical reassembly process [37–39] in which the adsorption process is instantaneously completed and followed by a subsequent slower transition from the amorphous to the crystalline phase in the presence of calcium ions only [39–42].

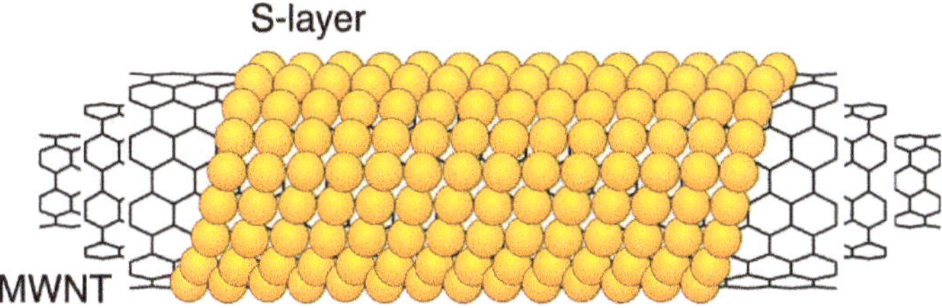

Figure 2. Schematic drawing of an S-layer coated multi walled carbon nanotube.

Finally, preliminary experiments with rSbpA$_{31-1068}$GG fusion protein and binding of colloidal gold labeled antibodies did not only give information about the orientation of the S-layer proteins bound on the carbon nanotubes but also about the functionality of the so coated and functionalized hybrid structures [29,30].

2. Materials and Methods

2.1. Production of Wild Type and Recombinant S-Layer Fusion Protein Solutions

L. sphaericus CCM 2177 (from the Czech Collection of Microorganisms) was grown in continuous culture as described in a previous study [43]. Cell wall fragments were obtained after a downstream process and used as starting point for the production of a monomeric wild type SbpA (wtSbpA) S-layer protein solution [44]. The protein was extracted with 5M guanidine hydrochloride (GHCL, Gerbu Nr. 1057) and after centrifugation dialyzed (membrane Biomol cut-off: 12–16 kD; pore size 2.5 nm) against 3 L Milli-Q water containing 2 mM ethylenediaminetetraacetic acid (EDTA). The resulting protein solution was adjusted to a final concentration of 1 mg/mL. The reassembly properties of the so obtained monomeric protein solution were determined by atomic force microscopy (AFM) (Multimode AFM, Bruker AXS, Santa Barbara, CA, USA) [45]. For this purpose, the protein solution was diluted with a crystallization buffer containing CaCl$_2$ to a final concentration of 100 µg/mL and applied on 10 × 10 mm sized silicon wafer pieces [39]. Only when the SbpA S-layer lattice with its characteristic square lattice symmetry could be visualized by AFM (Figure 1b), the protein solution was used for the experimental work and stored at 4 °C for a maximum of 4 weeks.

The S-layer fusion protein rSbpA$_{31-1068}$GG comprising a truncated—but still reassembly capable—form of SbpA with two IgG binding regions cloned from Protein G was expressed and purified as described in Reference [30]. The resulting monomeric protein solution was adjusted to a concentration of 1 mg/mL, the reassembly capability controlled by AFM, and stored as described above for the wtSbpA protein solution.

2.2. Coating of -COOH Functionalized MWNTs with S-Layer Proteins

In order to enhance the dispersion of multiwalled nanotubes (MWNTs) in aqueous buffer solutions carboxyl groups (-COOH) were introduced by oxidation following the protocol of Singer et al. in Reference [46]. Here, pristine MWNTs (SIGMA, Saint Louis, MO, USA; diameter of 50–90 nm, Nr. 901019) were suspended in 30% H_2O_2 (Roth, Nr. 8070) stirred and heated in an oil bath at 130 °C for a total of 4 h. After a filtration and washing step, the functionalized MWNTs were dried at 70 °C for 48 h.

These -COOH functionalized MWNTs were suspended in crystallization buffer (4 mg/30 mL; 5 mM Tris and 100 mM $CaCl_2$ in Milli-Q water, pH 9.0) under the aid of ultrasonication (Branson Sonifier 250; output 5, duty circle 50%) for 20 min. From this, still not well dispersed, solution 4.5 mL were transferred into a container with 500 µL wtSbpA or $rSbpA_{31\text{-}1068}GG$ monomeric protein solution (1 mg/mL) and ultrasonication was prolonged immediately for 4 min. This step was carried out in an ice bath to avoid possible denaturation of the S-layer protein caused by raising temperatures (typically > 45 °C) during ultrasonication. Subsequently, incubation was allowed to take place overnight at 4 °C using an overhead shaker (Heidolph, Reax 2, Schwabach, Germany). After 16 h the S-layer coated MWNTs were centrifuged (Eppendorf; Centrifuge 5424, Hamburg, Germany) at 5000 rcf for 10 min and resuspended in crystallization buffer to a final concentration of 1 mg/mL.

2.3. Immuno Gold Labeling of rSbpA₃₁₋₁₀₆₈GG Coated -COOH Functionalized MWNTs

From $rSbpA_{31\text{-}1068}GG$ coated MWNTs suspension 500 µL were centrifuged (Eppendorf; Centrifuge 5424) at 5000 rcf for 5 min, resuspended and incubated with gold labeled goat anti-human IgG (Amersham, AuroProbe™ BL plus, RPN 464F) 1:5 diluted in crystallization buffer (pH 9.0) containing 0.01% Triton-X 100 and 0.001% fish gelatin (Amersham, RPN 416V) at RT for 3 h. Goat IgG binds via the IgG binding moieties of the fusion protein. Subsequently, labeled MWNTs were washed once with crystallization buffer (centrifugation step: 5000 rcf for 10 min), negative stained, and investigated by transmission electron microscopy (TEM).

2.4. Transmission Electron Microscopy (TEM), Negative Staining, and Image Processing

The ability of S-layer proteins to form a crystalline and in the case of $rSbpA_{31\text{-}1068}GG$ a biologically active coating on MWNTs was demonstrated with an FEI Tecnai T20 Transmission Electron Microscope (TEM) operated at 160 kV (FEI Europe (now ThermoScientific), Eindhoven, The Netherlands) after negative staining of the samples. For this purpose, samples were adsorbed on 300 mesh copper grids (Christine Gröpl Elektronenmikroskopie, Tulln, Austria) coated with a Formvar-support film and a thin carbon layer. A chemical fixation of the protein adsorbed on the copper grids was done with a drop of 2.5% glutaraldehyde in cacodylate buffer (pH 7.4) for 10 min. For negative staining, samples were place on 2% uranium acetate drops for 10 min. All steps were performed at room temperature. The open source software ImageJ (version 1.52p) was used to straigthen bent S-layer coated MWNTs in TEM images and subsequently to calculate the Fourier spectrum showing the layer lines of the helical S-layer [47].

2.5. Preparation of Bucky Paper and Scanning Electron Microscopy

Buckypapers are simple membrane based CNT architectures [48]. In this study, buckypapers were prepared by filtration of wtSbpA coated MWNTs using an AMICON filtration unit (AMICON, Burlington, MA, USA; ultrafiltration cell Model 8010). A micro-filter (SARTORIUS, Goettingen, Germany; Sartolon polyamide, Nr. 25007-47-N) with a pore size of 0.2 µm was chosen as supporting membrane. After deposition of the S-layer functionalized MWNTs on the membrane surface, the buckpaper was removed from the filtration cell, air-dried and characterized with a ThermoScientific Apreo VS SEM (ThermoScientific, Eindhoven, The Netherlands) scanning electron microscope (SEM).

For this purpose, the bucky paper was cut into approx. 5 × 5 mm sized pieces and fixed with a conductive double sticky tape on standard 0.5″ aluminum stubs. The SEM was operated at 2.0 kV with a beam current of 0.1 nA in immersion (high resolution) mode. Images were recorded in high vacuum with the in-lens back-scattered electron (BSE) detector T1 and the secondary electron (SE) detector T2.

3. Results

3.1. Bucky Paper

Bucky paper was produced in order to make a quick check of the supplied MWNT diameters. The diameters were measured in the SEM images (n = 70) and found to be within the 50–90 nm range given by the manufacturer (70.1 ± 14.8nm; min = 43; max = 106 nm) (Figure 3). The bright areas in Figure 3b show the charging of the electrically insulating polymeric microfilter.

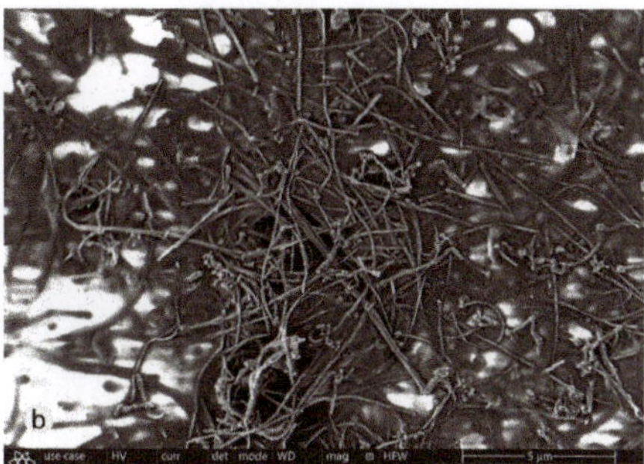

Figure 3. Bucky paper imaged in the SEM with the (**a**) BSE detector (T1) and the (**b**) SE detector (T2) detector. Note the difference between the material and topography contrast provided by the back-scattered and secondary electrons, respectively. Moreover, the pores in the microfiltration membrane are clearly visible in (**b**).

3.2. Coating of -COOH Functionalized MWNTs with wtSbpA S-Layer Protein

It was clear right from the beginning that a homogeneous dispersion of the MWNTs is absolutely necessary for the investigation and technological application of a successful S-layer coating. According to the literature, the MWNTs could not be easily dispersed in aqueous buffer solution even after oxidation with the associated introduction of -COOH groups and subsequent ultrasonication. This inhomogeneous suspension was not suitable for a successful S-layer coating, but when wtSbpA S-layer protein was added and ultrasonication prolonged, the suspension became immediately homogeneous (Figure 4). The suspension was stable then for at least several months (since the commencement of the work) and when required only had to be shaken in order to resuspend the sedimented S-layer coated MWNTs again.

It was assumed that the amphiphilic SbpA S-layer proteins were instantaneously attached to the MWNT surface and in this way shielded the highly hydrophobic MWNT surfaces from the aqueous medium. S-layer reassembly follows a two-stage non-classical pathway in which first extended monomers are attached to the surface, form amorphous and subsequently microcrystalline clusters from which crystalline order emerges by a final folding step. Calcium ions play an important role for the reassembly of most S-layer proteins including SbpA [39–42]. TEM investigations demonstrated that the S-layer completely covers the -COOH functionalized MWNTs (Figure 5).

Figure 4. Different MWNT suspensions after various consecutive treatments. (**a**) Resuspension of pristine MWNTs in recrystallization buffer. (**b**) Sample I: -COOH functionalized MWNTs in recrystallization buffer after ultrasonication (20 min). (**c**) Sample II: further treatment of sample I by adding wtSbpA S-layer protein in the course of a second short (4 min) ultrasonication step.

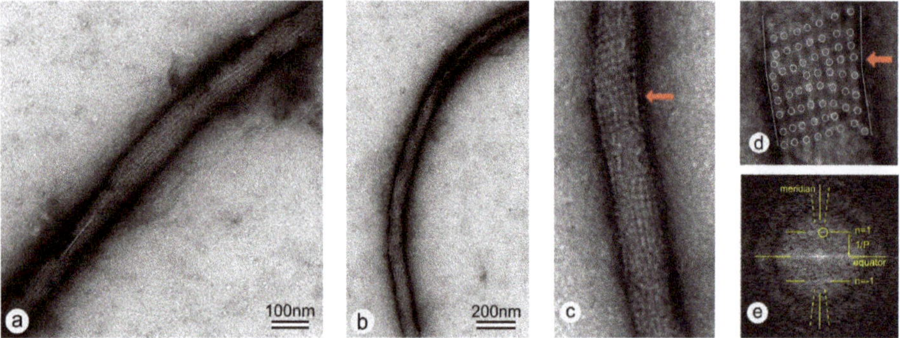

Figure 5. (**a**,**b**) TEM images of negatively stained wtSbpA coated MWNTs. The wtSbpA coating exhibits a crystalline lattice with square lattice symmetry. (**c**) Zoomed view of the lower end of the S-layer coated MWNT shown in (**b**). Note the lattice defect in the S-layer lattice (marked by the red arrow) where the MWNT buckles in (**c**) and in the further zoomed view in (**d**) where the unit cell positions are marked by circles too. (**e**) Fourier spectrum of a straightened copy of (**c**) showing the layer lines n = ±1 and in this way the corresponding pitch P of the helical S-layer.

A highly ordered protein layer exhibiting square lattice symmetry can be clearly seen, confirming the ability of wtSbpA to reassemble on -COOH functionalized MWNTs over long (several tens of micrometer) distances (Figure 6a). Lattice defects in the S-layer lattice, in particular disclinations, were found where the MWNT buckles (Figure 5c,d) [49]. Moreover and in general, the TEM image shows a side-on-view (elevation) of a helix with an axial repeat, termed pitch P. Thus, the Fourier spectrum will have an axial repeat along the Z-axis (the meridian) of 1/P. This generates a set of equally spaced layer lines separated by 1/P and indexed from Z = 0 (the equator) with n = 0, 1, 2, ... in the Z-direction and by negative integers in the –Z-direction. The amplitudes of the diffraction orders along a particular layer line are proportional to Bessel functions J_n of order n [50]. Because only J_0 is non-zero on the Z-axis, only the n = 0 layer line will be non-zero on the meridian. Successive first maxima of J_n progressively occur further from the Z-axis giving the appearance of a cross-characteristic for the Fourier spectrum of a helix- with intensities decreasing as n increases. In the Fourier spectrum (Figure 5e), only the first maximum in the first layer lines (n = ±1) is visible. The value of 1/P was determined with 1/13.67 nm which led us to the conclusion that the S-layer proteins with their unit cell size of 13.1 nm were arranged along a single basic helix or, in other words, that the helical repeat consisted of only one striation. The pitch angle was determined to be approximately 6.5°. For a more detailed description of how to analyze and index a diffraction pattern of a helical structure, see Reference [51]. Moreover, it has to be mentioned that the S-layer coating did not close the MWNT ends. This finding was not surprising since only S-layer proteins which reassemble in hexagonal (p6) lattice symmetry would be able to make rounded caps or closed vesicles [25] by the introduction of several lattice defects such as 5-fold wedge disclinations [52]. Although we have thought that the pitch of the S-layer lattice would be rather constant along the tube length, this assumption could not be confirmed in this work. This assumption was made since we know from previous work and the literature that the intrinsic curvature of the S-layer, as determined by the size of the respective bacterial cell, determines the curvature of self-assembly products in solution [53,54]. Moreover, it might also be possible that the chirality of the outer tube of the MWNTs influences the pitch of the S-layer, but previous studies with <100> and <111> silicon surfaces have shown that the S-layer did not resemble the silicon crystal structure because the ratio between the S-layer unit cell size in the 10 nm range and the lattice spacing of silicon in the 0.5 nm range was by far too large. This will be probably true for CNTs as well.

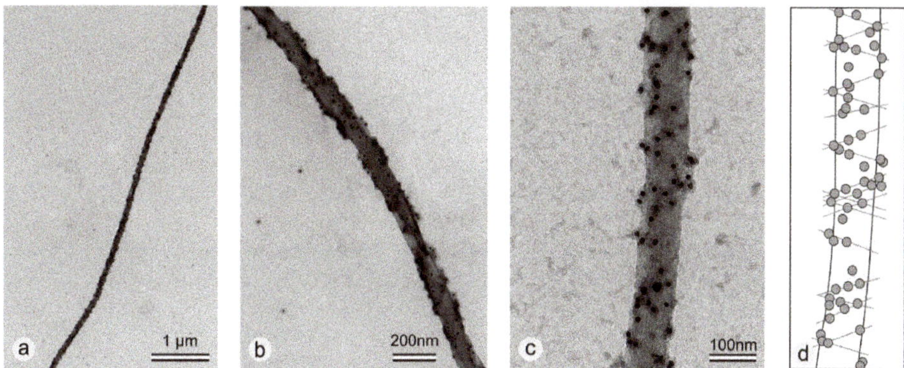

Figure 6. TEM images of immune gold labeled and negatively stained rSbpA$_{31-1068}$GG coated -COOH functionalized MWNTs. (**a**) The S-layer and in this was the gold nanoparticles cover tens of micrometers on the MWNT surface. (**b,c**) The gold particles of the gold-labeled goat antibodies can be clearly seen. It seems that they are helically arranged along the tubes as schematically shown in (**d**) for (**c**).

3.3. Coating of -COOH Functionalized MWNTs with rSbpA$_{31-1068}$GG Fusion Protein

In addition to the investigation of the wild type S-layer protein (wtSbpA), -COOH functionalized MWNTs were coated with the recombinant S-layer fusion protein rSbpA$_{31-1068}$GG comprising the IgG binding region of Protein G. It could be demonstrated that also the recombinant S-layer protein coating resulted in a homogeneous distribution of the MWNTs within the buffer solution. Immune gold labeling and TEM were used to prove the general concept of functionalizing MWNTs with a tailor-made highly specific S-layer fusion protein. Gold nanoparticles, which were bound to the IgG moieties, could only be seen at the rSbpA$_{31-1068}$GG coated MWNT samples (Figure 6) while in blank experiments gold nanoparticles were not found on wtSbpA coated or uncoated MWNTs (data not shown). Moreover, it appears that the gold nanoparticles resembled the helical arrangement of the S-layer along the MWNTs (Figure 6c,d).

Although the crystalline lattice structure was not clearly visible, the immune gold labeling of the rSbpA$_{31-1068}$GG coated MWNTs confirmed the functionality of the coating. Moreover, since the IgG binding moieties were introduced at the C-terminus of the fusion protein and, in general, binding to solid supports is favored via the N-terminus, it was concluded that the rSbpA$_{31-1068}$GG was attached with its N-terminus while the C-terminus presenting the IgG binding moieties was directed towards the outside surrounding medium [29].

3.4. Dispersion of -COOH Functionalized MWNTs by Addition of Triton-X 100

It has to be mentioned that we have also tried to increase the dispersibility of MWNTs by adding Triton-X 100 which immediately led to homogeneous suspensions [8]. However, it is also known that

Triton-X 100 interferes with the reassembly properties of S-layer proteins. Optimization of the protocol by lowering the Triton-X 100 concentration (below 0.01%, best with 0.003% in crystallization buffer) allowed the dispersion of -COOH functionalized MWNTs in aqueous buffer solution and—at a first glance—did not hinder the reassembly of the S-layer proteins on solid supports and MWNTs (see Supplemental Material, Figure S1). However, unfortunately, the recrystallization of wtSbpA did not occur on the surface of the MWNTs but mainly in the form of self-assembly sheets attached to the MWNTs (see Supplemental Material, Figure S2). It could not be clarified whether these self-assembly products start growing from attached S-layer proteins into the surrounding medium or whether they were detached from a loose sheathing. Therefore, the described protocol for the dispersion and S-layer coating of -COOH functionalized MWNTs—starting with ultrasonication in recrystallization buffer containing $CaCl_2$ for a total of 20 min, subsequently adding S-layer protein and continuing ultrasonication for further 4 min—was established as the standard protocol for this and future work.

4. Discussion

Besides chemical modifications [55], coating with synthetic polymers or surfactants [56,57], or DNA [57], also coating with proteins is seen as a promising but challenging technique to disperse and functionalize CNTs [5,9,13,36]. In general, non-covalent approaches are favored as they preserve the properties of CNTs while improving their dispersibility. The usage of BSA, DNA [58], hydrophobins [13,36], and lysozyme, which was able to disperse coated CNTs in a pH dependent way [59], was already reported. In addition, encapsulation by proteins makes CNTs not only more biocompatible but also less toxic [7,36] and offers the advantage that new functional hybrid structures with the beneficial properties of both may be developed [60]. The high surface area of CNTs and their ability to pass cell membranes make them ideal vehicles to transport drugs into cells [61,62]. Nevertheless, it has to be considered that the functionality of bound proteins may be impaired due to their random adsorption and denaturation on the CNT surface [63].

Thus, we would like to stress that the S-layer and carbon nanotube construction kit which is introduced in this work will offer the advantage to generate entirely new carriers and containers when used as catalysts, templates, scaffolds, or affinity matrices. For example, we have already shown that S-layer fusion proteins with particularly tailored bio-reactive domains allow a highly specific and sensitive (unsurpassed) functionalization of surfaces in the development of biosensor surfaces (for review see Reference [22,24]). We will make use of this knowledge and specifically bind biomolecules on native and genetically functionalized S-layer fusion protein coated MWNTs (Figure 7a). CNTs and graphene have already shown great potential in the development of biosensors due to their huge surface area, great electron transfer rate, good electrical conductivity, and ability to immobilize biologically functional molecules, such as enzymes, aptamers, or receptors [12–19]. Concerning S-layers, a considerable amount of knowledge has accumulated concerning the fabrication of amperometric sensors for glucose [64] or sucrose [65] or of optical sensors for glucose too [66]. Moreover, based on our experience in fabricating multi-enzyme amperometric biosensors, it will be possible to develop stoichiometrically well-adjusted multi-layer CNT-supported sensing layers too [65]. A promising new approach for the production of biocatalysts comprises the use of S-layer lattices that present functional multimeric enzymes on their surface, thereby forming a most accurate spatial distribution, orientation, and stability of these enzymes [67]. In comparison to conventional approaches, S-layers provide a biocompatible surface endowed with the capability to bind the functional biomolecules in a dense packing. Our approach is not limited to the use of S-layers as binding matrices only. The key feature of S-layer fusion proteins is their functional domain such as the IgG binding domain (shown in this work) [29–31], the Bet-v1 domain specific to the major birch pollen allergen [68], or for a broad range of applications the biotin binding domain [69] and affinity tag for streptavidin [70] (for review see References [22,24]). Moreover, it has to be stressed that the successful labeling of the IgG binding moieties allowed to unambiguously conclude that the S-layer was oriented with its outer face towards the medium and the biological functionality maintained—a basic requirement in our developments. In addition, S-layer fusion proteins will allow us to perform our studies on a broader scale including material sciences when, for example, fluorescent S-layer fusion proteins with fluorescent domains (EGFP, ECFP, YFP, RFP1) or FRET pairs (Förster- or Fluorescent-resonance energy transfer) [71–73] are used. A generic approach for fluorescent proteins may be found in Reference [74]. S-layer fusion protein lattices will allow a much more specific and versatile functionalization of CNTs (Figure 7a) [69] compared to the classical approach where biomolecules are bound more-or-less randomly through diversely located functional groups (Figure 7b) [5].

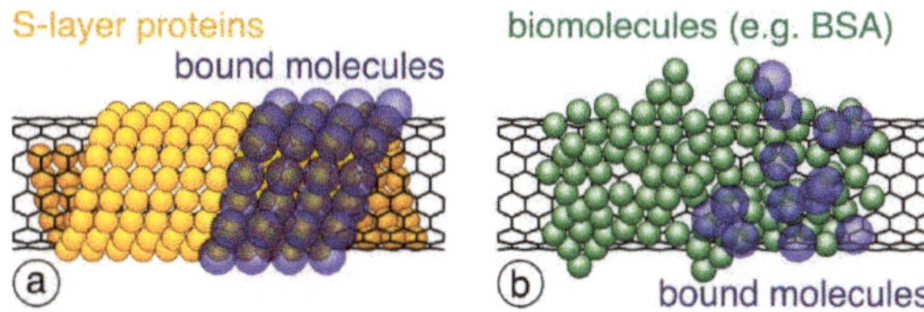

Figure 7. Schematic drawing of (**a**) precisely arranged S-layer proteins (shown in orange) around the CNT surface and (**b**) randomly bound biomolecules (e.g., BSA; shown in green). When used for binding further biomolecules (shown in blue), only S-layers provide a dense and precisely controlled packing.

It has to be stressed also that S-layers act as stabilizing and tethering structures and ionic reservoirs for biomembranes [75] and might provide an advantage in the development of a biomembrane based field effect transistor (FET) [76].

Moreover, tuning of the anti-fouling properties of the S-layer coating might be a further advantage for particular applications [36,77]. Alternatively, when using S-layer fusion proteins with catalytic sites exposed, a highly efficient templated biomineralization of metallic, semiconducting, or other inorganic materials will be facilitated. This might be particularly interesting for fundamental research in solid state physics, when an S-layer lattice is used as template for generating an ordered metallic nanoparticle array [78–84] directly on the CNT surface [85].

In the course of this work, it turned out that the S-layer coating increases the dispersibility of MWNTs in water dramatically, and it may be anticipated that their biocompatibility will be improved and at the same time their cytotoxicity reduced [36,57,86]. Moreover, since our TEM investigations demonstrated a good long-range order of the S-layer along the nanotubes over several (tens of) micrometers, it may be assumed that the non-classical multi stage reassembly pathway of S-layer proteins might be the key for the defect-free lattice formation over such large distances too [37–39]. Contrary to the classical approach, the multi stage process allows the healing of lattice defects in the growing crystalline domains and is pre-requisite for a self-purifying effect in the course of lattice formation [87]. Based on work with hydrophobic silicon surfaces, we assume that the healing step must be particularly favored on the highly hydrophobic surface of CNTs [45].

5. Summary

We would like to stress that our research, although longer term in nature, will provide technologies and materials which are more versatile to conventional approaches in the development of functional surfaces in terms of sensitivity, selectivity, and density of functional groups. Although the usage of protein coatings (e.g., BSA, lysozyme or hydrophobins) to disperse carbon nanotubes in aqueous solutions is well described in the literature, the added value of native and functionalized S-layer fusion proteins is based on their unique reassembly properties and, in this context, precisely aligned functional groups and domains for binding additional bioactive molecules and compounds.

Supplementary Materials: The following are available online at http://www.mdpi.com/2079-6412/9/8/492/s1, Figure S1: AFM image of wtSbpA recrystallized on a silicon wafer surface in the presence of 0.01% Triton-X 100. Although no complete protein layer could be observed, the square lattice could be easily detected, confirming the recrystallization properties in the presence of low Triton-X 100 concentrations in crystallization buffer. Figure S2:

TEM images of negatively stained wtSbpA coated MWNTs. The MWNTs had been dispersed in 0.01% Triton-X 100 before the addition of wtSbpA. The S-layer protein showed the ability to recrystallize but not by coating the MWNTs themselves but in the form of attached self-assembly products exhibiting the square lattice symmetry.

Author Contributions: Conceptualization, D.P., A.B., and U.B.S.; methodology, D.P., A.B., P.S., H.L., and U.B.S.; validation, D.P., A.B., and U.B.S.; formal analysis, D.P., A.B., and U.B.S.; investigation, D.P., A.B., P.S., H.L., and U.B.S.; resources, D.P.; data curation, D.P. and A.B.; writing—original draft preparation, D.P. and A.B.; writing—review and editing, D.P., A.B., P.S., H.L., and U.B.S.; visualization, D.P. and A.B.; supervision, D.P.; project administration, D.P.; funding acquisition, D.P.

Funding: Part of this research was funded by the Austrian Science Fund (FWF) [project P 31927-N28].

Acknowledgments: We would like to thank Max Willinger for his support in the preparation of MWNTs.

Conflicts of Interest: The authors declare no conflict of interest.

References

1. Iijima, S. Helical Microtubules of Graphitic Carbon. *Nature* **1991**, *354*, 56–58. [CrossRef]
2. Dresselhaus, M.S.; Dresselhaus, G.; Saito, R. Physics of Carbon Nanotubes. *Carbon* **1995**, *33*, 883–891. [CrossRef]
3. Avouris, P.; Chen, Z.H.; Perebeinos, V. Carbon-based electronics. *Nat. Nanotechnol.* **2007**, *2*, 605–615. [CrossRef] [PubMed]
4. Saeed, K.; Khan, I. Carbon nanotubes-properties and applications: A review. *Carbon Lett.* **2013**, *14*, 131–144. [CrossRef]
5. Nagaraju, K.; Reddy, R.; Reddy, N. A review on protein functionalized carbon nanotubes. *J. Appl. Biomater. Func.* **2015**, *13*, E301–E312. [CrossRef]
6. Hirsch, A. Functionalization of single-walled carbon nanotubes. *Angew. Chem. Int. Ed.* **2002**, *41*, 1853–1859. [CrossRef]
7. Balasubramanian, K.; Burghard, M. Chemically functionalized carbon nanotubes. *Small* **2005**, *1*, 180–192. [CrossRef]
8. Kharissova, O.V.; Kharisov, B.I.; de Casas Ortiz, E.G. Dispersion of carbon nanotubes in water and non-aqueous solvents. *RSC Adv.* **2013**, *3*, 24812–24852. [CrossRef]
9. Calvaresi, M.; Zerbetto, F. The Devil and Holy Water: Protein and Carbon Nanotube Hybrids. *Acc. Chem. Res.* **2013**, *46*, 2454–2463. [CrossRef]
10. Posseckardt, J.; Zhang, J.W.; Mertig, M. Mobility of a supported lipid bilayer on dispersed single-walled carbon nanotubes. *Phys. Status Solidi A* **2016**, *213*, 1427–1433. [CrossRef]
11. Haft, M.; Gronke, M.; Gellesch, M.; Wurmehl, S.; Buchner, B.; Mertig, M.; Hampel, S. Tailored nanoparticles and wires of Sn, Ge and Pb inside carbon nanotubes. *Carbon* **2016**, *101*, 352–360. [CrossRef]
12. Jacobs, C.B.; Peairs, M.J.; Venton, B.J. Review: Carbon nanotube based electrochemical sensors for biomolecules. *Anal. Chim. Acta* **2010**, *662*, 105–127. [CrossRef] [PubMed]
13. Wang, X.S.; Wang, H.C.; Huang, Y.J.; Zhao, Z.X.; Qin, X.; Wang, Y.Y.; Miao, Z.Y.; Chen, Q.A.; Qiao, M.Q. Noncovalently functionalized multi-wall carbon nanotubes in aqueous solution using the hydrophobin HFBI and their electroanalytical application. *Biosens. Bioelectron.* **2010**, *26*, 1104–1108. [CrossRef] [PubMed]
14. Vashist, S.K.; Zheng, D.; Al-Rubeaan, K.; Luong, J.H.T.; Sheu, F.S. Advances in carbon nanotube based electrochemical sensors for bioanalytical applications. *Biotechnol. Adv.* **2011**, *29*, 169–188. [CrossRef] [PubMed]
15. Liu, S.; Guo, X.F. Carbon nanomaterials field-effect-transistor-based biosensors. *NPG Asia Mater.* **2012**, *4*, 1–10. [CrossRef]
16. Gomes, F.O.; Maia, L.B.; Delerue-Matos, C.; Moura, I.; Moura, J.J.G.; Morais, S. Third-generation electrochemical biosensor based on nitric oxide reductase immobilized in a multiwalled carbon nanotubes/1-n-butyl-3-methylimidazolium tetrafluoroborate nanocomposite for nitric oxide detection. *Sens. Actuators B Chem.* **2019**, *285*, 445–452. [CrossRef]
17. Kumar, S.; Bukkitgar, S.D.; Singh, S.; Singh, V.; Reddy, K.R.; Shetti, N.P.; Reddy, C.V.; Sadhu, V.; Naveen, S. Electrochemical Sensors and Biosensors Based on Graphene Functionalized with Metal Oxide Nanostructures for Healthcare Applications. *Chemistryselect* **2019**, *4*, 5322–5337. [CrossRef]

18. Kwon, O.S.; Song, H.S.; Park, T.H.; Jang, J. Conducting Nanomaterial Sensor Using Natural Receptors. *Chem. Rev.* **2019**, *119*, 36–93. [CrossRef]

19. Wayu, M.B.; Pannell, M.J.; Labban, N.; Case, W.S.; Pollock, J.A.; Leopold, M.C. Functionalized carbon nanotube adsorption interfaces for electron transfer studies of galactose oxidase. *Bioelectrochemistry* **2019**, *125*, 116–126. [CrossRef]

20. Pantarotto, D.; Singh, R.; McCarthy, D.; Erhardt, M.; Briand, J.P.; Prato, M.; Kostarelos, K.; Bianco, A. Functionalized carbon nanotubes for plasmid DNA gene delivery. *Angew. Chem. Int. Ed.* **2004**, *43*, 5242–5246. [CrossRef]

21. Jovanovic, S.P.; Markovic, Z.M.; Kleut, D.N.; Romcevic, N.Z.; Trajkovic, V.S.; Dramicanin, M.D.; Markovic, B.M.T. A novel method for the functionalization of gamma-irradiated single wall carbon nanotubes with DNA. *Nanotechnology* **2009**, *20*. [CrossRef] [PubMed]

22. Sleytr, U.B.; Schuster, B.; Egelseer, E.M.; Pum, D. S-layers: Principles and applications. *FEMS Microbiol. Rev.* **2014**, *38*, 823–864. [CrossRef] [PubMed]

23. Pum, D.; Sleytr, U.B. Reassembly of S-layer proteins. *Nanotechnology* **2014**, *25*, 312001. [CrossRef] [PubMed]

24. Egelseer, E.M.; Ilk, N.; Pum, D.; Messner, P.; Schäffer, C.; Schuster, B.; Sleytr, U.B. S-Layers, microbial, biotechnological applications. In *Encyclopedia of Industrial Biotechnology: Bioprocess, Bioseparation, and Cell Technology*; Flickinger, M.C., Ed.; John Wiley and Sons: Hoboken, NJ, USA, 2010; Volume 7, pp. 4424–4448.

25. Sleytr, U.B. Self-assembly of the hexagonally and tetragonally arranged subunits of bacterial surface layers and their reattachment to cell walls. *J. Ultrastruct. Res.* **1976**, *55*, 360–377. [CrossRef]

26. Ladenhauf, E.M.; Pum, D.; Wastl, D.S.; Toca-Herrera, J.L.; Phan, N.V.H.; Lieberzeit, P.A.; Sleytr, U.B. S-layer based biomolecular imprinting. *RSC Adv.* **2015**, *5*, 83558–83564. [CrossRef]

27. Schuster, D.; Kupcu, S.; Belton, D.J.; Perry, C.C.; Stoger-Pollach, M.; Sleytr, U.B.; Pum, D. Construction of silica-enhanced S-layer protein cages. *Acta Biomater.* **2013**, *9*, 5689–5697. [CrossRef]

28. Ilk, N.; Egelseer, E.M.; Sleytr, U.B. S-layer fusion proteins—Construction principles and applications. *Curr. Opin. Biotechnol.* **2011**, *22*, 824–831. [CrossRef]

29. Völlenkle, C.; Weigert, S.; Ilk, N.; Egelseer, E.; Weber, V.; Loth, F.; Falkenhagen, D.; Sleytr, U.B.; Sara, M. Construction of a functional S-layer fusion protein comprising an immunoglobulin G-binding domain for development of specific adsorbents for extracorporeal blood purification. *Appl. Environ. Microbiol.* **2004**, *70*, 1514–1521. [CrossRef]

30. Ucisik, M.H.; Küpcü, S.; Breitwieser, A.; Gelbmann, N.; Schuster, B.; Sleytr, U.B. S-layer fusion protein as a tool functionalizing emulsomes and CurcuEmulsomes for antibody binding and targeting. *Colloids Surf. B* **2015**, *128*, 132–139. [CrossRef]

31. Breitwieser, A.; Pum, D.; Toca-Herrera, J.L.; Sleytr, B.U. Magnetic beads functionalized with recombinant S-layer protein exhibit high human IgG-binding and anti-fouling properties. *Curr. Top. Pept. Protein Res.* **2016**, *17*, 45–55.

32. Scheicher, S.R.; Kainz, B.; Kostler, S.; Reitinger, N.; Steiner, N.; Ditlbacher, H.; Leitner, A.; Pum, D.; Sleytr, U.B.; Ribitsch, V. 2D crystalline protein layers as immobilization matrices for the development of DNA microarrays. *Biosens. Bioelectron.* **2013**, *40*, 32–37. [CrossRef]

33. Pum, D.; Sleytr, U.B. Large-scale reconstruction of crystalline bacterial surface layer proteins at the air-water interface and on lipids. *Thin Solid Films* **1994**, *244*, 882–886. [CrossRef]

34. Ilk, N.; Vollenkle, C.; Egelseer, E.M.; Breitwieser, A.; Sleytr, U.B.; Sara, M. Molecular characterization of the S-layer gene, sbpA, of Bacillus sphaericus CCM 2177 and production of a functional S-layer fusion protein with the ability to recrystallize in a defined orientation while presenting the fused allergen. *Appl. Environ. Microb.* **2002**, *68*, 3251–3260. [CrossRef]

35. Pavkov-Keller, T.; Howorka, S.; Keller, W. The structure of bacterial S-layer proteins. *Prog. Mol. Biol. Transl. Sci.* **2011**, *103*, 73–130. [CrossRef]

36. Yang, W.; Ren, Q.; Wu, Y.N.; Morris, V.K.; Rey, A.A.; Braet, F.; Kwan, A.H.; Sunde, M. Surface functionalization of carbon nanomaterials by self-assembling hydrophobin proteins. *Biopolymers* **2013**, *99*, 84–94. [CrossRef]

37. Chung, S.; Shin, S.H.; Bertozzi, C.R.; De Yoreo, J.J. Self-catalyzed growth of S layers via an amorphous-to-crystalline transition limited by folding kinetics. *Proc. Natl. Acad. Sci. USA* **2010**, *107*, 16536–16541. [CrossRef]

38. Shin, S.H.; Chung, S.; Sanii, B.; Comolli, L.R.; Bertozzi, C.R.; De Yoreo, J.J. Direct observation of kinetic traps associated with structural transformations leading to multiple pathways of S-layer assembly. *Proc. Natl. Acad. Sci. USA* **2012**, *109*, 12968–12973. [CrossRef]

39. Breitwieser, A.; Iturri, J.; Toca-Herrera, J.L.; Sleytr, U.B.; Pum, D. In Vitro Characterization of the Two-Stage Non-Classical Reassembly Pathway of S-Layers. *Int. J. Mol. Sci.* **2017**, *18*, 400. [CrossRef]

40. Pum, D.; Sleytr, U.B. Anisotropic crystal growth of the S-layer of Bacillus sphaericus CCM 2177 at the air/water interface. *Colloids Surf. A* **1995**, *102*, 99–104. [CrossRef]

41. Baranova, E.; Fronzes, R.; Garcia-Pino, A.; Van Gerven, N.; Papapostolou, D.; Pehau-Arnaudet, G.; Pardon, E.; Steyaert, J.; Howorka, S.; Remaut, H. SbsB structure and lattice reconstruction unveil Ca^{2+} triggered S-layer assembly. *Nature* **2012**, *487*, 119–122. [CrossRef]

42. Rad, B.; Haxton, T.K.; Shon, A.; Shin, S.H.; Whitelam, S.; Ajo-Franklin, C.M. Ion-specific control of the self-assembly dynamics of a nanostructured protein lattice. *ACS Nano* **2015**, *9*, 180–190. [CrossRef] [PubMed]

43. Ilk, N.; Kosma, P.; Puchberger, M.; Egelseer, E.M.; Mayer, H.F.; Sleytr, U.B.; Sára, M. Structural and functional analyses of the secondary cell wall polymer of *Bacillus sphaericus* CCM 2177 that serves as an S-layer-specific anchor. *J. Bacteriol.* **1999**, *181*, 7643–7646. [PubMed]

44. Egelseer, E.M.; Leitner, K.; Jarosch, M.; Hotzy, C.; Zayni, S.; Sleytr, U.B.; Sára, M. The S-layer proteins of two *Bacillus stearothermophilus* wild-type strains are bound via their N-terminal region to a secondary cell wall polymer of identical chemical composition. *J. Bacteriol.* **1998**, *180*, 1488–1495. [PubMed]

45. Györvary, E.S.; Stein, O.; Pum, D.; Sleytr, U.B. Self-assembly and recrystallization of bacterial S-layer proteins at silicon supports imaged in real time by atomic force microscopy. *J. Microsc.* **2003**, *212*, 300–306. [CrossRef] [PubMed]

46. Singer, G.; Siedlaczek, P.; Sinn, G.; Rennhofer, H.; Micusik, M.; Omastova, M.; Unterlass, M.M.; Wendrinsky, J.; Milotti, V.; Fedi, F.; et al. Acid Free Oxidation and Simple Dispersion Method of MWCNT for High-Performance CFRP. *Nanomaterials* **2018**, *8*, 912. [CrossRef] [PubMed]

47. Schneider, C.A.; Rasband, W.S.; Eliceiri, K.W. NIH Image to ImageJ: 25 years of image analysis. *Nat. Methods* **2012**, *9*, 671–675. [CrossRef]

48. Yang, X.S.; Lee, J.; Yuan, L.X.; Chae, S.R.; Peterson, V.K.; Minett, A.I.; Yin, Y.B.; Harris, A.T. Removal of natural organic matter in water using functionalised carbon nanotube buckypaper. *Carbon* **2013**, *59*, 160–166. [CrossRef]

49. Harris, W.F. Disclinations. *Sci. Am.* **1977**, *237*, 130–145. [CrossRef]

50. DeRosier, D.J.; Moore, P.B. Reconstruction of 3-Dimensional Images from Electron Micrographs of Structures with Helical Symmetry. *J. Mol. Biol.* **1970**, *52*, 355–369. [CrossRef]

51. Diaz, R.; Rice, W.J.; Stokes, D.L. Fourier-Bessel Reconstruction of Helical Assemblies. *Methods Enzymol.* **2010**, *482*, 131–165. [CrossRef]

52. Messner, P.; Pum, D.; Sára, M.; Stetter, K.O.; Sleytr, U.B. Ultrastructure of the cell envelope of the archaebacteria Thermoproteus tenax and Thermoproteus neutrophilus. *J. Bacteriol.* **1986**, *166*, 1046–1054. [CrossRef] [PubMed]

53. Messner, P.; Pum, D.; Sleytr, U.B. Characterization of the ultrastructure and the self-assembly of the surface layer of Bacillus stearothermophilus strain NRS 2004/3a. *J. Ultrastruct. Mol. Struct. Res.* **1986**, *97*, 73–88. [CrossRef]

54. Bobeth, M.; Blecha, A.; Bluher, A.; Mertig, M.; Korkmaz, N.; Ostermann, K.; Rodel, G.; Pompe, W. Formation of tubes during self-assembly of bacterial surface layers. *Langmuir* **2011**, *27*, 15102–15111. [CrossRef] [PubMed]

55. Ali-Boucetta, H.; Nunes, A.; Sainz, R.; Herrero, M.A.; Tian, B.; Prato, M.; Bianco, A.; Kostarelos, K. Asbestos-like pathogenicity of long carbon nanotubes alleviated by chemical functionalization. *Angew. Chem. Int. Ed.* **2013**, *52*, 2274–2278. [CrossRef] [PubMed]

56. Rastogi, R.; Kaushal, R.; Tripathi, S.K.; Sharma, A.L.; Kaur, I.; Bharadwaj, L.M. Comparative study of carbon nanotube dispersion using surfactants. *J. Colloid Interface Sci.* **2008**, *328*, 421–428. [CrossRef] [PubMed]

57. Mallakpour, S.; Soltanian, S. Surface functionalization of carbon nanotubes: Fabrication and applications. *RSC Adv.* **2016**, *6*, 109916–109935. [CrossRef]

58. Awasthi, K.; Singh, D.P.; Singh, S.; Dash, D.; Srivastava, O.N. Attachment of biomolecules (protein and DNA) to amino-functionalized carbon nanotubes. *New Carbon Mater.* **2009**, *24*, 301–306. [CrossRef]

59. Nepal, D.; Geckeler, K.E. pH-sensitive dispersion and debundling of single-walled carbon nanotubes: Lysozyme as a tool. *Small* **2006**, *2*, 406–412. [CrossRef]

60. Zhang, Y.J.; Li, J.; Shen, Y.F.; Wang, M.J.; Li, J.H. Poly-L-lysine functionalization of single-walled carbon nanotubes. *J. Phys. Chem. B* **2004**, *108*, 15343–15346. [CrossRef]

61. Liu, Z.; Robinson, J.T.; Tabakman, S.M.; Yang, K.; Dai, H.J. Carbon materials for drug delivery & cancer therapy. *Mater. Today* **2011**, *14*, 316–323. [CrossRef]

62. Lacerda, L.; Raffa, S.; Prato, M.; Bianco, A.; Kostarelos, K. Cell-penetrating CNTs for delivery of therapeutics. *Nano Today* **2007**, *2*, 38–43. [CrossRef]

63. Butler, J.E.; Ni, L.; Brown, W.R.; Joshi, K.S.; Chang, J.; Rosenberg, B.; Voss, E.W., Jr. The immunochemistry of sandwich ELISAs–VI. Greater than 90% of monoclonal and 75% of polyclonal anti-fluorescyl capture antibodies (CAbs) are denatured by passive adsorption. *Mol. Immunol.* **1993**, *30*, 1165–1175. [CrossRef]

64. Neubauer, A.; Pum, D.; Sleytr, U.B. An Amperometric Glucose Sensor-Based on Isoporous Crystalline Protein Membranes as Immobilization Matrix. *Anal. Lett.* **1993**, *26*, 1347–1360. [CrossRef]

65. Neubauer, A.; Hodl, C.; Pum, D.; Sleytr, U.B. A Multistep Enzyme Sensor for Sucrose Based on S-Layer Microparticles as Immobilization Matrix. *Anal. Lett.* **1994**, *27*, 849–865. [CrossRef]

66. Neubauer, A.; Pum, D.; Sleytr, U.B.; Klimant, I.; Wolfbeis, O.S. Fibre-optic glucose biosensor using enzyme membranes with 2-D crystalline structure. *Biosens. Bioelectron.* **1996**, *11*, 317–325. [CrossRef]

67. Ferner-Ortner-Bleckmann, J.; Gelbmann, N.; Tesarz, M.; Egelseer, E.M.; Sleytr, U.B. Surface-layer lattices as patterning element for multimeric extremozymes. *Small* **2013**, *9*, 3887–3894. [CrossRef] [PubMed]

68. Breitwieser, A.; Egelseer, E.M.; Moll, D.; Ilk, N.; Hotzy, C.; Bohle, B.; Ebner, C.; Sleytr, U.B.; Sara, M. A recombinant bacterial cell surface (S-layer)-major birch pollen allergen-fusion protein (rSbsC/Bet v1) maintains the ability to self-assemble into regularly structured monomolecular lattices and the functionality of the allergen. *Protein Eng.* **2002**, *15*, 243–249. [CrossRef]

69. Moll, D.; Huber, C.; Schlegel, B.; Pum, D.; Sleytr, U.B.; Sara, M. S-layer-streptavidin fusion proteins as template for nanopatterned molecular arrays. *Proc. Natl. Acad. Sci. USA* **2002**, *99*, 14646–14651. [CrossRef]

70. Huber, C.; Liu, J.; Egelseer, E.M.; Moll, D.; Knoll, W.; Sleytr, U.B.; Sara, M. Heterotetramers formed by an S-layer-streptavidin fusion protein and core-streptavidin as a nanoarrayed template for biochip development. *Small* **2006**, *2*, 142–150. [CrossRef]

71. Kainz, B.; Steiner, K.; Moller, M.; Pum, D.; Schaffer, C.; Sleytr, U.B.; Toca-Herrera, J.L. Absorption, steady-state fluorescence, fluorescence lifetime, and 2D self-assembly properties of engineered fluorescent S-layer fusion proteins of Geobacillus stearothermophilus NRS 2004/3a. *Biomacromolecules* **2010**, *11*, 207–214. [CrossRef]

72. Kainz, B.; Steiner, K.; Sleytr, U.B.; Pum, D.; Toca-Herrera, J.L. Fluorescence energy transfer in the bi-fluorescent S-layer tandem fusion protein ECFP-SgsE-YFP. *J. Struct. Biol.* **2010**, *172*, 276–283. [CrossRef] [PubMed]

73. Kainz, B.; Steiner, K.; Sleytr, U.B.; Pum, D.; Toca-Herrera, J.L. Fluorescent S-layer protein colloids. *Soft Matter* **2010**, *6*, 3809–3814. [CrossRef]

74. Dunakey, S.J.G.; Coyle, B.L.; Thomas, A.; Xu, M.; Swift, B.J.F.; Baneyx, F. Selective Labeling and Decoration of the Ends and Sidewalls of Single-Walled Carbon Nanotubes Using Mono- and Bispecific Solid-Binding Fluorescent Proteins. *Bioconjugate Chem.* **2019**, *30*, 959–965. [CrossRef] [PubMed]

75. Schuster, B.; Sleytr, U.B. Composite S-layer lipid structures. *J. Struct. Biol.* **2009**, *168*, 207–216. [CrossRef]

76. Gong, H.; Chen, F.; Huang, Z.L.; Gu, Y.; Zhang, Q.Z.; Chen, Y.J.; Zhang, Y.; Zhuang, J.; Cho, Y.K.; Fang, R.N.H.; et al. Biomembrane-Modified Field Effect Transistors for Sensitive and Quantitative Detection of Biological Toxins and Pathogens. *ACS Nano* **2019**, *13*, 3714–3722. [CrossRef]

77. Rothbauer, M.; Küpcü, S.; Sticker, D.; Sleytr, U.B.; Ertl, P. Exploitation of S-layer Anisotropy: pH-dependent Nanolayer Orientation for Cellular Micropatterning. *ACS Nano* **2013**, *7*, 8020–8030. [CrossRef]

78. Shenton, W.; Pum, D.; Sleytr, U.B.; Mann, S. Biocrystal templating of CdS superlattices using self-assembled bacterial S-layers. *Nature* **1997**, *389*, 585–587. [CrossRef]

79. Dieluweit, S.; Pum, D.; Sleytr, U.B. Formation of a gold superlattice on an S-layer with square lattice symmetry. *Supramol. Sci.* **1998**, *5*, 15–19. [CrossRef]

80. Mertig, M.; Kirsch, R.; Pompe, W.; Engelhardt, H. Fabrication of highly oriented nanocluster arrays by biomolecular templating. *Eur. Phys. J. D* **1999**, *9*, 45–48. [CrossRef]

81. Pompe, W.; Mertig, M.; Kirsch, R.; Wahl, R.; Ciacchi, L.C.; Richter, J.; Seidel, R.; Vinzelberg, H. Formation of metallic nanostructures on biomolecular templates. *Zeitschrift Fur Metallkunde* **1999**, *90*, 1085–1091.

82. Mertig, M.; Wahl, R.; Lehmann, M.; Simon, P.; Pompe, W. Formation and manipulation of regular metallic nanoparticle arrays on bacterial surface layers: An advanced TEM study. *Eur. Phys. J. D* **2001**, *16*, 317–320. [CrossRef]

83. Queitsch, U.; Mohn, E.; Schaffel, F.; Schultz, L.; Rellinghaus, B.; Bluher, A.; Mertig, M. Regular arrangement of nanoparticles from the gas phase on bacterial surface-protein layers. *Appl. Phys. Lett.* **2007**, *90*, 113114. [CrossRef]

84. Mann, S. Self-assembly and transformation of hybrid nano-objects and nanostructures under equilibrium and non-equilibrium conditions. *Nat. Mater.* **2009**, *8*, 781–792. [CrossRef]

85. Grigoryan, G.; Kim, Y.H.; Acharya, R.; Axelrod, K.; Jain, R.M.; Willis, L.; Drndic, M.; Kikkawa, J.M.; DeGrado, W.F. Computational design of virus-like protein assemblies on carbon nanotube surfaces. *Science* **2011**, *332*, 1071–1076. [CrossRef]

86. Vardharajula, S.; Ali, S.Z.; Tiwari, P.M.; Eroglu, E.; Vig, K.; Dennis, V.A.; Singh, S.R. Functionalized carbon nanotubes: Biomedical applications. *Int. J. Nanomed.* **2012**, *7*, 5361–5374. [CrossRef]

87. Sleutel, M.; Van Driessche, A.E.S. Role of clusters in nonclassical nucleation and growth of protein crystals. *Proc. Natl. Acad. Sci. USA* **2014**, *111*, E546–E553. [CrossRef]

 *coatings*

Article

Physical and Morphological Characterization of Chitosan/Montmorillonite Films Incorporated with Ginger Essential Oil

Victor Gomes Lauriano Souza [1,*], João Ricardo Afonso Pires [1] Carolina Rodrigues [1], Patricia Freitas Rodrigues [2], Andréia Lopes [2], Rui Jorge Silva [2], Jorge Caldeira [3,4,5], Maria Paula Duarte [1], Francisco Braz Fernandes [2], Isabel Maria Coelhoso [3] and Ana Luisa Fernando [1,*]

[1] MEtRICs, Departamento de Ciências e Tecnologia da Biomassa, Faculdade de Ciências e Tecnologia, FCT, Universidade NOVA de Lisboa, Campus de Caparica, 2829-516 Caparica, Portugal; jr.pires@campus.fct.unl.pt (J.R.A.P.); cpe.rodrigues@campus.fct.unl.pt (C.R.); mpcd@fct.unl.pt (M.P.D.)

[2] CENIMAT/I3N, Departamento de Ciência dos Materiais, Faculdade de Ciências e Tecnologia, Universidade NOVA de Lisboa, Campus de Caparica, 2829-516 Caparica, Portugal; pf.rodrigues@campus.fct.unl.pt (P.F.R.); keyalopes@gmail.com (A.L.); rjcs@fct.unl.pt (R.J.S.); fbf@fct.unl.pt (F.B.F.)

[3] LAQV-REQUIMTE, Departamento de Química, Faculdade de Ciências e Tecnologia, Universidade Nova de Lisboa, 2829-516, Campus de Caparica, 2829-516 Caparica, Portugal; fjfc@fct.unl.pt (J.C.); imrc@fct.unl.pt (I.M.C.)

[4] UCIBIO-REQUIMTE, Departamento de Química, Faculdade de Ciências e Tecnologia, Universidade Nova de Lisboa, 2829-516, Campus de Caparica, 2829-516 Caparica, Portugal

[5] Centro de investigação interdisciplinar Egas Moniz, Instituto Universitário Egas Moniz, Campus Universitário, Quinta da Granja, Monte de Caparica, 2829 - 511 Caparica, Portugal

[*] Correspondence: v.souza@campus.fct.unl.pt (V.G.L.S.); ala@fct.unl.pt (A.L.F.)

Received: 11 October 2019; Accepted: 24 October 2019; Published: 26 October 2019

Abstract: Novel bionanocomposite films of chitosan/montmorillonite (CS/MMT) activated with ginger essential oil (GEO) were produced and characterized in terms of their physical and morphological properties. The homogenization process led to a good interaction between the chitosan and the nanoparticles, however the exfoliation was diminished when GEO was incorporated. Film glass transition temperature did not statistically change with the incorporation of either MMT or GEO, however the value was slightly reduced, representing a relaxation in the polymer chain which corroborated with the mechanical and barrier properties results. Pristine chitosan films showed excellent barrier properties to oxygen with a permeability of 0.184×10^{-16} mol/m·s·Pa being reduced to half (0.098×10^{-16} mol/m·s·Pa) when MMT was incorporated. Although the incorporation of GEO increased the permeability values to 0.325×10^{-16} mol/m·s·Pa when 2% of GEO was integrated, this increment was smaller with both MMT and GEO (0.285×10^{-16} mol/m·s·Pa). Bionanocomposites also increased the UV light barrier. Thus, the produced bioplastics demonstrated their ability to retard oxidative processes due to their good barrier properties, corroborating previous results that have shown their potential in the preservation of foods with high unsaturated fat content.

Keywords: biobased polymers; chitosan; nanotechnology

---

1. Introduction

The development of food packaging is currently focused on the use of polymers usually derived from plants, i.e., bioplastics with an eco-friendly and sustainable approach, an alternative to non-biodegradable materials from non-renewable sources (petroleum-based materials) [1–4]. These materials can be divided into four main classes, namely: proteins, polysaccharides, lipids and composites [1]. Polysaccharides are highlighted due to their good oxygen barrier related to their well-ordered hydrogen bonded network shape, which may enhance the protection conferred to food packaged against the oxidation process [1,5,6]. However, the weaker mechanical properties and water vapor barrier limit their use, and through the combination with other polymers (i.e., bilayers) or the incorporation of nanofillers, such as montmorillonite, this limitation can be overcome [5,7,8]. Chitosan (poly-β(1,4)-2-amino-2-deoxy-D-glucose), the second most abundant polysaccharide in nature, obtained from the deacetylation of chitin, is a biodegradable biopolymer with potential to be use as food-grade films and coatings [9–16].

Aiming to reduce the use of synthetic chemical additives, the food industry has increased its interest in the research of natural preservatives, i.e., food components or extracts with antimicrobial and antioxidant properties, with less harmful effect to human health [17–20]. Moreover, food-borne microbial outbreaks have driven the industry and the scientific community to search for innovative ways to inhibit microbial growth in foods while maintaining quality, freshness, and safety [21]. The use of packaging is an option to provide an increased margin of safety and quality, as the next generation of food packaging will include materials with antimicrobial and antioxidant properties that are capable of protecting the food packaged, extending its shelf life [7,21,22]. Essential oils (EOs) are a good example of such materials suitable to be used in the production of this next generation of food packaging [23]. EOs are natural substances extracted from different parts (i.e., roots, bark, or leaves) of a variety of plants, many of them already used in traditional culinary (e.g., rosemary, thyme, sage, ginger, citronella) [18]. Despite the remarkable preservative properties of these components, their use as food additives is discouraged by the strong aroma they confer to the foodstuff when directly incorporated into the food matrix [17,18,22,24]. Thus, the incorporation of EOs into food packaging material may solve this organoleptic problem, enabling their application to the preservation of food [24]. When incorporated into polymeric matrices, the amount necessary is diminished as the migration occurs gradually towards the food surface, where the food spoilage takes place [25,26].

In our previous works, different natural extracts, either essential oils or hydro-alcoholic extracts from a variety of plants, were incorporated into chitosan [17,27]. The films produced were characterized in terms of their antioxidant activity [17] and physical properties [27]. Within the different essential oils tested, it was concluded that rosemary essential oil (REO) and ginger essential oil (GEO) presented the best active properties [17], thus, novel bionanocomposites incorporated with such EOs were produced and nanoreinforced with montmorillonite (MMT). As novel materials, it is crucial to understand how the incorporation of MMT and the essential oil would interfere in the functional and bioactive properties of the film. In vitro and in situ activity of CS/MMT activated with different levels of GEO have been investigated [28] and those bionanocomposites have proven to have good antimicrobial and antioxidant activity, being able to extend the shelf-life of fresh poultry meat, a foodstuff with high unsaturated fat content. As the ultimate purpose of the film is to be used as primary packaging for food products, information on its mechanical and barrier properties, not studied yet for these novel materials, should also be understood. Thus, this work is focused on the characterization of bionanocomposites incorporated with GEO reinforced with MMT in terms of their physical, morphological, and barrier properties.

2. Materials and Methods

2.1. Materials and Reagents

Commercial high molecular weight (31–37 kDa) chitosan (poly(D-glucosamine)) with 75% of deacetylation and ethanol absolute were purchased from Sigma Aldrich (Germany). Sodium montmorillonite (Cloisite®Na⁺) was kindly supplied by BYK Additives & Instruments (USA). Ginger essential oil (EO), with food grade classification, was acquired from Biover (Belgium). Glacial acetic acid, glycerol, calcium nitrate (Ca(NO$_3$)$_2$), sodium bromide (NaBr), potassium acetate (CH$_3$COOK, 99% purity), and tween 80 (polyethylene glycol sorbitan monolaurate) were purchased from Alfa Aesar (Germany). Sodium chloride

(NaCl) was obtained from PanReac (Spain). All chemicals were of analytical reagent grade and were used as purchased. The water was purified using the Milli-Q system (Millipore, USA).

2.2. Bionanocomposites Production

The composite films were prepared according to Souza et al. (2018) [28]. Briefly, to prepare the film-forming dispersion (FFD), 1.5% (*w/v*) of chitosan was dissolved in 1% (*v/v*) of glacial acetic acid solution under continuous agitation for 24 h at room temperature. Then, 30% (*w/w* chitosan) of glycerol was added as a plasticizer in all treatments. At this stage, GEO in the levels tested (0%; 0.5%; 1% or 2% *v/v* FFD) and 0.2% (*w/v* in essential oil) of the emulsifier tween 80 were added to the system. Subsequently, an agitation cycle consisting of 5 min agitation with ultraturrax (15,000 rpm) (IKA®T18, Staufen, Germany) followed by 15 min degasification in an ultrasound bath (360 W) (Selecta, Spain) was carried out. The resulting dispersion was then casted in glass molds (18 × 25 cm) and let to dry naturally for approximately 48–72 h. This was the procedure used to produce chitosan films without the incorporation of the nanoreinforcement (MMT). To produce the bionanocomposites, 2.5% (*w/w* chitosan) of MMT was added to the FFD already containing the glycerol, and two extra agitation steps (same as described before) were added before the incorporation of GEO and tween 80, and the third and final agitation cycle was carried out, followed by the casting. These extras steps were added to supply energy to the system and promote the exfoliation of the MMT. Dried films were peeled and stored protected from light in a desiccator containing saturated calcium nitrate solution at 25 °C and 50% relative humidity, monitored with a thermohygrometer, until evaluation.

2.3. Film Characterization

2.3.1. X-ray Diffraction (XRD)

X-ray diffraction is one of the techniques used to study the structure of MMT within the polymer, being indicative of a succeeded exfoliation process, thus a tool to understand whether the nanomaterial reinforced the biopolymer or not. Diffractograms of the films were obtained using a DMAX-IIIC diffractometer (Rigaku Industrial Corporation, Tokyo, Japan), equipped with CuKα (λ = 1.5418 Å) radiation (40 kV, 30 mA), 2θ angle range 5° to 40°, a scanning rate of 2°/min and a sampling interval of 0.02° (2θ). The interlamellar distances (d001) were calculated using Bragg's Law (Equation (1)) [22].

$$sen\ \theta = \frac{n\lambda}{2d} \tag{1}$$

where λ is the radiation wavelength, d is the interlamellar distance in Å, n is diffraction number (n = 1), and θ is the measured diffraction angle.

2.3.2. Attenuated Total Reflectance Fourier Transform Infrared Spectroscopy (ATR-FTIR)

Attenuated total reflectance Fourier transform infrared spectra of the bionanocomposites were collected using a FTIR spectrometer (model PerkinElmer spectrum Two, Perkin Elmer, Waltham, MA, USA) from 4000 to 650 cm^{-1} at a 1 cm^{-1} resolution [17].

2.3.3. Morphological Characterization: Scanning Electron Microscopy (SEM)

To analyze the bionanocomposite morphology, scanning electron microscopy micrographs were performed from the surface and cross-section of the following samples: pristine chitosan film, chitosan + MMT; chitosan + 2% GEO, and chitosan + MMT + 2% GEO. The images were obtained from a Zeiss instrument (Model DSM 962, Oberkochen, Germany) under vacuum, accelerated at 3 kV. The samples were fixed with a double adhesive coated carbon tape on aluminium stubs and covered with gold palladium using a sputter coater. Cross-section images were carried out in samples previously submitted to fractions after contact with liquid nitrogen.

2.3.4. Thermal Properties

Thermal properties were studied through differential scanning calorimetry (DSC) and thermogravimetric analysis (TGA).

The glass transition temperature (Tg) of the samples was determined using DSC model 204 F1 Phoenix® (Netzsch, Selb, Germany). Approximately 10 mg of each sample were sealed in a standard aluminum pan and thermal cycle was performed from 20 °C to 350 °C at a constant heating rate of 20 °C/min under nitrogen atmosphere (flow rate of 50 mL/min) [29].

The thermal stability of bionanocomposites was characterized using a simultaneous thermal analyzer (PerkinElmer, Model STA 6000, Germany). A sample from each treatment (approximately 10 mg) was heated to 900 °C at a rate of 10 °C/min and maintained in isotherm during 3 min under nitrogen atmosphere [30]. Compound degradation temperatures and percentages were determined from the first derivative of the weight loss curve percent (DTGA) versus temperature.

2.3.5. Thickness and Mechanical Properties

Film thickness was measured using a Mitutoyo digital micrometer (Mitutoyo, Kawasaki, Japan), with 0.001 mm precision, on ten randomly points of each sample [22].

Mechanical properties were determined according to ASTM D882–12 (2012) [31]. Elastic modulus (EM), tensile strength (TS), and percentage of elongation at break (EAB) were measured from the tensile testing of five strips of each film with dimensions 150 mm wide and 25.4 mm long. The samples were mounted in the tensile grips with a 0.5 kN load cell (Autograph Shimadzu, Sydney, Australia), with 50 mm initial gauge length and stretched at a cross-head speed of 50 mm/min until breakage.

2.3.6. Optical Properties

To calculate optical parameters chroma (c*) and Hue angle (hue), CIE-L*a*b* coordinates (L* indicates black (0) to white (100); a* indicates red (+) to green (−) and b* indicates yellow (+) to blue (−)) were measured from the bionanocomposites using a colorimeter CR 410 (Minolta Co., Tokyo, Japan) with a 10 mm diameter window and D65 illuminant/10° observer. The measurements were taken on a white background standard and the equations used are the following [32]:

$$c^* = \left(a^{*2} + b^{*2}\right)^{1/2} \tag{2}$$

$$hue = \arctan\left(\frac{b^*}{a^*}\right)x\frac{180}{\pi}, \text{ for } a^* > 0 \text{ and } b^* > 0 \tag{3}$$

$$hue = \arctan\left(\frac{b^*}{a^*}\right)x\frac{180}{\pi} + 180, \text{ for } a^* < 0 \tag{4}$$

$$hue = \arctan\left(\frac{b^*}{a^*}\right)x\frac{180}{\pi} + 360, \text{ for } a^* > 0 \text{ and } b^* < 0 \tag{5}$$

Film opacity was also determined by direct reading of the absorbance of rectangular samples at 600 nm using a UV–vis spectrophotometer (Model Spekol 1500, Analytikjena, Germany) and calculated according to Equation (6) [33].

$$Opacity\left(mm^{-1}\right) = \frac{absorbance\ 600\ nm}{sample\ thickness\ (mm)} \tag{6}$$

Finally, film transparency was obtained from spectrum scans (from 190 to 900 nm) using a UV–vis spectrophotometer of each film specimen. Air was used as reference and the results were expressed as a percentage of transmittance [34].

2.3.7. Contact Angle (CA)

Contact angle (CA) was measured using a goniometer (KSV Instruments Ltd., CAM 100, Finland) with the software KSV CAM 100 at room temperature (25 ± 2 °C) on the bionanocomposites produced. Film hydrophilic character was evaluated from water drop contact angles with their upper surface [5].

2.3.8. Solubility and Swelling Degree

The solubility in water (g/100 g of film) and the swelling degree (g/100 g of film) were determined according to Souza et al. (2017) [27]. Briefly, film specimens were cut into a rectangle (2×2 cm) and weighted (precision 0.0001 g) in an analytical balance (Mettler Toledo AB204, Switzerland), obtaining the initial weight (M1); then samples were dried at 70 °C for 24 h in a natural conventional oven (WTB binder, Germany), and were weighted to obtain the initial dry mass (M2). Subsequently, samples were placed in Petri dishes containing 30 mL of Milli-Q water and stored for 24 h at room temperature (25 ± 2 °C), in order to allow the swelling process. After this contact period, the specimens were superficially dried with filter paper and weighted (M3) again. The residual film specimens were dried in an oven at 70 °C for 24 h to determine the final dry mass (M4). Two measurements from each film sample were taken, and the parameters calculate according to Equations (7) and (8).

$$\% \ Solubility \left(\frac{g}{100g} \ of \ film \right) = \frac{(M_2 - M_4)}{M_2} \times 100 \tag{7}$$

$$\% \ Swelling \ degree \left(\frac{g}{100g} \ of \ film \right) = \frac{(M_3 - M_2)}{M_2} \times 100 \tag{8}$$

2.3.9. Water Vapor Permeability (WVP)

The WVP (mol/m·s·Pa) was determined gravimetrically at 30 °C, based on the method described by Ferreira et al. (2016) [5]. The tested films were sealed on the top of 45 mm diameter glass cells containing 8 mL of saturated NaCl solution (relative humidity (RH) = 76.9%) and placed in a desiccator with saturated potassium acetate solution (RH = 22.5 %) equipped with a fan to promote air circulation and maintain constant driving force. Temperature and the relative humidity were monitored with a thermohygrometer (Vaisala, Finland). The water transferred through the film and absorbed by the desiccant was determined from weight loss of the permeation cell (measured every 1 h during 10 h), and the WVP calculated by following equation (Equation (9)):

$$WVP = \frac{N_W \times \delta}{\Delta P_{w.eff}} \tag{9}$$

where Nw (mol/m^2·s) is the water vapour flux, δ (m) is the film thickness and $\Delta P_{w.eff}$ (Pa) is the effective driving force. Results are the average ± standard deviation of the three replicates analyzed.

2.3.10. Oxygen Permeability (OP)

The OP was determined in a stainless steel cell with two identical chambers separated by the tested film [5,35]. Tested films were previously equilibrated at 30 °C and relative humidity of 55% ± 5% (desiccator containing saturated sodium bromide solution). The OP was assessed by pressurizing one of the chambers (feed) up to 0.7 bar with pure oxygen (99.999% purity) (Praxair, Spain), followed by the measurement of the pressure change in both chambers over time, using two pressure transducers (Jumo, Model 404327, Germany). The system was kept inside a thermostatic water bath at 30 °C (Julabo, Model EH, Germany). The permeability was calculated using Equation (10):

$$\frac{1}{\beta} \ln \left(\frac{\Delta p_0}{\Delta p} \right) = P \frac{t}{\delta} \tag{10}$$

where Δp (mbar) is the pressure difference between the feed and permeate compartment, P (mol·m/m^2·s·Pa) is the gas permeability, t (s) is the time, δ (m) is the film thickness and β is the geometric parameter of the cell [35].

2.4. Data Statistical Treatment

All experiments were conducted using a completely randomized design with three replications.

A statistical analysis of data was performed through a one-way analysis of variance using OriginLab software version 8.5, and differences among mean values were processed by the Tukey test. Significance was defined at $p < 0.05$.

3. Results and Discussion

3.1. X-ray Diffraction

The diffractograms of pristine MMT and pure chitosan films or films incorporated with sodium montmorillonite and/or with ginger essential oil are shown in Figure 1. Montmorillonite exhibited a characteristic reflection peak at about $2\theta = 7.512°$, corresponding to a basal spacing between the individual MMT layers of d001 = 1.18 nm. In the film incorporated only with MMT, the nanoclay characteristic reflection peak has disappeared, probably as the result of a disordered configuration of the lamellar structure not detectable by XRD, providing strong evidence that the clay nanolayers are exfoliated [36,37]. The incorporation of GEO resulted in the shift of the characteristic montmorillonite crystalline peak to smaller angles around $2\theta = 5.02°$ (0.5% GEO), 5.18° (1% GEO) and 5.22° (2% GEO), corresponding to a basal spacing between the individual MMT layers of d001 = 1.76; 1.71, and 1.69 nm respectively. Similar values (2θ between 5.04° and 5.20°), were found for the interplanar distance in chitosan films incorporated with rosemary essential oil [16]. The increase of the distances between the lamellae of the clay is an indication that the chitosan was able to intercalate between the MMT layers, obtaining an intercalated structure [36,38]. However, ginger essential oil reduced chitosan dispersion in the MMT galleries, probably due to structural changes in the polymer due to the interactions of the phenolic compounds with the chitosan reactive groups [37].

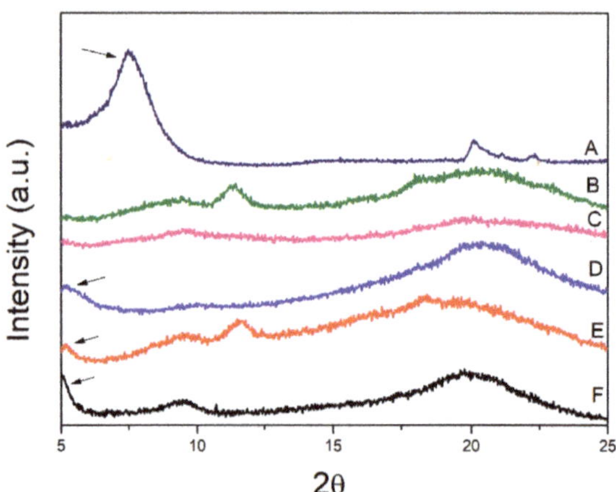

Figure 1. XRD of (**A**) pristine sodium montmorillonite (MMT), (**B**) chitosan film, (**C**) chitosan + MMT film, (**D**) chitosan + MMT + 2% GEO film, (**E**) chitosan + MMT + 1% GEO film, and (**F**) chitosan + MMT + 0.5% GEO film.

3.2. Attenuated Total Reflectance Fourier Transform Infrared Spectroscopy (ATR-FTIR)

Figure 2 depicts the FT-IR spectra of the bionanocomposites produced. Chitosan characteristic absorption bands were observed in all spectra, namely: at 3325 cm^{-1} (axial stretch of –OH); at 3265–3277 cm^{-1} (asymmetric stretch of the –NH group); at 2877–2925 cm^{-1} (C–H bond of the methyl group –NHCOCH$_3$); at 1638–1642 cm^{-1} (amide I); at 1551–1558 cm^{-1} (amide II); at 1342 cm^{-1} (skeletal vibration involving the stretching of the C–N bond of amide III); at 1375–1412 cm^{-1} (–CH2 folding); at 906–1024 cm^{-1} (skeletal vibration involving the stretching of the C–O group); and at 1134 cm^{-1} (asymmetric stretching of the C–O–C bridges) [29,39,40]. Overall, the incorporation of GEO or MMT did not result in great differences in the spectra when compared to pristine chitosan film, probably due to the small quantity incorporated, i.e., chitosan characteristic peaks have prevailed in all samples. However, small changes in the intensities of the absorption peaks were recorded, which are attributed to the overlap of chemical bonds, and thus an indication of the presence of strong interaction between the molecules of the different components of the material (chitosan, MMT, and the active compounds present in GEO) [41]. The spectra of the films incorporated with GEO show a new peak between 1702–1703 cm^{-1} which corresponds to the vibration of the C=O bond stretch, the increase in this peak intensity with increasing oil concentration is an indication that there were interactions between the phenolic compounds present in the essential oil and the hydroxyl and amine groups of chitosan [42]. The appearance of this new peak as well as the displacement of peaks at different wavelengths (as observed between 860–880 cm^{-1} or 1250–1300 cm^{-1}, for example) with the incorporation of GEO are indicative that new covalent bonds between the chitosan and active compounds from GEO or MMT occurred [43]. These results corroborate the observations obtained in the analyses of XRD and SEM as well as the modifications in the functional properties of the films produced.

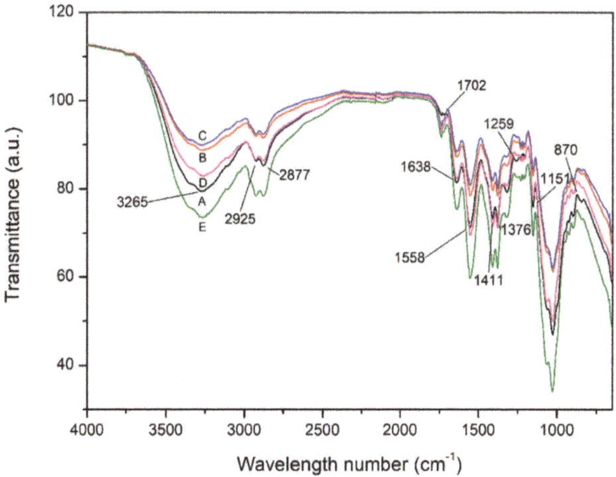

Figure 2. ATR-FTIR spectra of chitosan (Ch)-based films (**A**) incorporated with: (**B**) MMT; (**C**) MMT + 0.5% GEO; (**D**) MMT + 1% GEO; and (**E**) MMT + 2% GEO.

3.3. Morphological Characterization

Scanning electron microscopy images in Figure 3 correspond to the surface (Figure 3A,C,E,G) and cross-section (Figure 3B,D,F,G) of chitosan films and those containing MMT, 2% GEO, or 2% GEO + MMT.

The surface of the film showed high uniformity (Figure 3). However, the addition of GEO (2%) caused a certain discontinuity in the topography of the films (Figure 3E), which was more evident when MMT was also incorporated (Figure 3G). This less homogeneous surface (greater irregularity/less smooth) may have been the result of the presence of droplets under the surface of the film (internal aggregates) that changed the topography of the film.

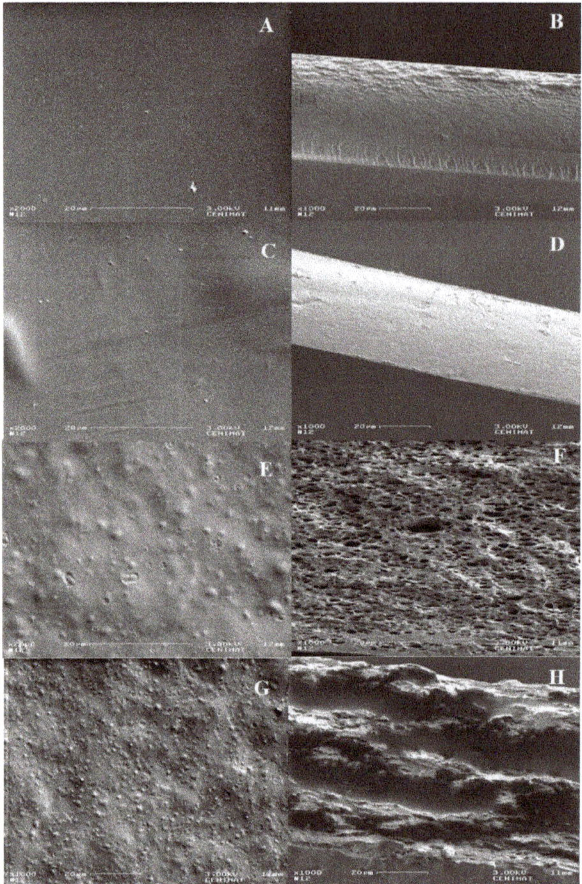

Figure 3. Scanning electron microscopy micrographs of surface and cross-section, respectively: chitosan film (**A** and **B**), chitosan film + MMT (**C** and **D**), chitosan film + 2% GEO (**E** and **F**), and chitosan film + MMT + 2% GEO (**G** and **H**).

The cross-sections of the pristine chitosan film (Figure 3B) and chitosan + MMT (Figure 3D) show high homogeneity of the polymer structure was obtained. The incorporation of MMT seems to have contributed to make the network even more compact, being an indicative of achievement of high interaction between chitosan and the nanoclay, which corroborates the XRD results. The incorporation of 2% of GEO resulted in an internal structure with a "spongy like" pattern (Figure 3F), probably due to the presence of internal oil droplets that also increases the surface coarseness [44]. When both oil and MMT were incorporated (Figure 3H), there was a complete change in the network structure (with an increase in the internal spacing), however with indications of greater interaction than in the film only incorporated with GEO.

Figure 3F,H shows the cross section of chitosan + 2% GEO and chitosan + 2% GEO + MMT, respectively; they are in the same amplification used for the images of pristine chitosan film (Figure 3B) and chitosan + MMT (Figure 3D); however, it is noticeable that the formers have wider cross sections (larger images). This observation is related to the increase in the thickness resulting from the incorporation of the essential oil, which corroborates the results discussed in Section 3.5.

3.4. Thermal Properties

The glass transition temperatures as well as the mass losses with their respective maximum degradation temperatures were determined by the DSC and TGA assays, respectively.

The Tg of the films did not statistically change with the incorporation of MMT nor GEO ($p > 0.05$) (Table 1), with an average value of 199 ± 14 °C. However, there was a tendency for Tg to decrease with the incorporation of GEO and MMT, which is an indication of a greater relaxation of the polymer chains, probably due to the plasticizing effect of the essential oil [45]. Similar results were found for chitosan films incorporated with sodium montmorillonite and rosemary essential oil [16]. Moreover, this tendency also corroborates the mechanical results discussed in Section 3.5.

Thermogravimetric analysis is a technique where the mass of a substance is monitored as a function of temperature increase under controlled conditions of temperature and atmosphere (usually under an inert gas flow such as nitrogen) [46]. The thermal degradation temperatures (Td), mass losses (% ΔM) and the sample residues (%) are shown in Table 1.

Regarding the thermal degradation process, two different behaviors were observed, namely: for the films without incorporation of GEO, three thermal degradation events were registered, whereas for biopolymers incorporated with GEO, four stages were found. Similar behavior was observed in chitosan films incorporated with *Satureja hortensis* essential oil [14].

The first thermal event occurred in average temperatures between 59.6 °C and 70.9 °C with mass losses between 4.1% and 6.5%. The mass loss in this first stage is related to the evaporation of the water and residual acetic acid present in the polymer matrix. The lower mass loss observed for the films incorporated with GEO was probably due to the lower water content resulting from the incorporation of the hydrophobic compounds present in the essential oil [47]. The second thermal degradation stage (135.5 °C–183.4 °C) corresponded to a loss of mass between 11.1% and 16.6% and is related to the decomposition of low molecular weight or structurally bonded components to water in the chitosan network [14].

The third, and largest mass loss (19.9%–33%) occurred at temperatures between 283.5 °C–290.8 °C and is associated with the degradation of chitosan (the main component of the bionanocomposites), i.e., it is associated with the dehydration of the saccharide ring, depolymerization, and pyrolytic decomposition of acetylated or deacetylated chitosan units [14,36]. The incorporation of GEO slightly increased the thermal stability of the film since the maximum degradation temperature at this stage was higher in films incorporated with the essential oil. These results are in good agreement with those observed in chitosan-carboxymethyl cellulose films which were also incorporated with ginger essential oil [48]. Those authors attributed the increase in thermal stability to the increase in the organization of the polymer matrix (i.e., to a more homogeneous structure) with the incorporation of GEO, which resulted in higher temperatures of thermal degradation [48].

The fourth and final stage of degradation occurred between 397.1 °C and 409.5 °C, and only for the films incorporated with GEO. This stage is related to the degradation of the thermally stable compounds present in the GEO, as also observed by Alizadeh et al. (2018) [14].

The incorporation of MMT acted as a thermal barrier, providing lower mass losses, as can be observed in the final residue of the thermal process. The increase in thermal stability induced by the addition of clays to polymer composites is commonly observed and is related to the level of dispersion and the aspect ratio obtained [13]. Exfoliation/intercalation of the clay layers between the polymer matrix increases the tortuosity of the combustion gases diffusion pathway, favoring the formation of a protection on the surface of the material (thermal insulation), contributing to the increase in degradation temperatures [11,13]. This finding is in agreement with a previous study [49].

The thermal decomposition residues of biopolymers at the end of the heating cycle (up to 900 °C) incorporated with GEO were lower compared to the control, whereas the composites with MMT increased the amount of ash. Similar behaviors were reported in the literature—Alizadeh et al. (2018) [14] observed the decrease of mineral residue with the incorporation of essential oil in chitosan films, whereas Rimdusit et al. (2008) [50] reported a slight increase in the amount of ash in the methylcellulose biopolymers incorporated with montmorillonite and attributed this increase to the inorganic clay characteristic.

Table 1. Summary of the thermal analyses results of active bionanocomposites.

Film	Tg (°C)	Δ1		Δ2		Δ3		Δ4		Residue (%)
		Td (°C)	ΔM (%)	Td (°C)	ΔM (%)	Td (°C)	ΔM (%)	Td (°C)	ΔM (%)	
Ch	206.7 ± 3.7 [ns**]	70.9 ± 2.5	6.5 ± 0.0	170.1 ± 1.5	12.6 ± 0.8	283.6 ± 0.7	24.3 ± 0.6	–	–	23.6 ± 0.4
Ch + MMT	188.9 ± 6.7 [ns]	63.6 ± 2.4	6.4 ± 0.8	173.2 ± 0.8	14.0 ± 0.8	283.5 ± 0.5	23.2 ± 0.2	–	–	25.8 ± 0.1
Ch + 0.5% GEO	202.2 ± 2.7 [ns]	65.6 ± 2.4	6.1 ± 0.2	174.1 ± 2.1	13.7 ± 0.1	284.0 ± 1.0	20.4 ± 0.4	407.4 ± 5.6	26.6 ± 0.6	18.0 ± 0.1
Ch+MMT + 0.5% GEO	202.8 ± 5.4 [ns]	64.0 ± 0.0	6.2 ± 0.3	168.4 ± 3.2	11.5 ± 0.6	284.0 ± 0.9	20.4 ± 0.0	407.8 ± 0.4	26.9 ± 0.4	20.2 ± 0.4
Ch + 1% GEO	194.4 ± 7.1 [ns]	63.2 ± 1.5	4.9 ± 0.2	167.4 ± 1.3	11.1 ± 2.7	284.1 ± 2.5	19.9 ± 1.7	397.1 ± 6.2	27.1 ± 5.7	16.9 ± 0.8
Ch + MMT + 1% GEO	196.2 ± 3.4 [ns]	64.8 ± 1.9	4.7 ± 0.1	172.6 ± 3.7	16.1 ± 0.3	290.8 ± 1.1	25.2 ± 0.4	409.5 ± 2.6	26.9 ± 0.8	17.7 ± 0.3
Ch + 2% GEO	195.5 ± 4.8 [ns]	59.6 ± 6.3	4.1 ± 0.2	135.5 ± 3.2	13.1 ± 1.3	284.8 ± 0.7	33.0 ± 0.5	407.1 ± 3.2	25.0 ± 0.6	14.2 ± 1.0
Ch + MMT + 2% GEO	194.3 ± 6.8 [ns]	63.7 ± 0.9	5.0 ± 0.2	183.4 ± 2.5	16.6 ± 0.6	287.7 ± 0.5	23.2 ± 0.9	408.7 ± 3.1	25.9 ± 0.6	18.0 ± 0.4

Chitosan (Ch); sodium montmorillonite (MMT); ginger essential oil (GEO); glass transition temperature (Tg); decomposition temperature (Td); mass loss (ΔM). ** ns: non-significance statistical difference ($p > 0.05$).

3.5. Thickness and Mechanical Properties

The incorporation of MMT did not statistically change the film thickness ($p > 0.05$; Table 2). However, in general, the thickness of the films incorporated with MMT were slightly lower than those found for the films without nanoreinforcement, and in some cases this difference was statistically significant (chitosan + MMT + 0.5% GEO or 2% GEO) ($p < 0.05$). As verified in the XRD assay, the films incorporated with MMT and GEO showed evidence of obtaining an intercalated (partially exfoliated) configuration, characterized by the formation of strong bonds between the polymer and the clay due to penetration of the chitosan chains [51], allowing the formation of a compact structure which minimized the increase in thickness due to the incorporation of GEO.

On the other hand, the incorporation of GEO resulted in a significant increase in the film thickness; moreover, a concentration effect was also observed, i.e., the greater the GEO content incorporated, the greater the thickness of the samples ($p < 0.05$; Table 2). This behavior can be explained by the higher content of solids per unit area [52], or due to the interactions between the chitosan and the active compounds present in the bioactive extracts that may have reduced the alignment of the polymer chains, reducing the compression of the formed network [47] and consequently increasing the thickness of the films.

Peng and Li (2014) [53] observed an increase in the thickness of chitosan films due to the incorporation of 1% of three different essential oils. Pure chitosan film had an average thickness of 77 μm whereas in treatments incorporated with citronella, thyme, or cinnamon essential oils, the recorded thicknesses were 97, 101, and 99 μm, respectively. That is, an increase of about 30% compared to the control. In our results, for the same essential oil content, the films showed an increase in thickness of about 62%. In another study, also with chitosan films incorporated with a mixture of ginger and cinnamon essential oils (1:1) in different concentrations (0.05%, 0.2%, or 1%), the thickness of the films produced varied significantly due to the addition of EOs (between 68 and 105 μm) [54]. This variation represents a 54% increase in the thickness with the addition of 1% of the mixture of the oils used, which is a result close to the 62% found in the present work.

The most common parameters that describe the mechanical properties of edible films are the maximum tensile strength at break (TS), the elongation at break (%EAB) and the elastic or Young's modulus (EM), which are strongly related to the chemical structure of the material. The TS indicates the film strength, %EAB corresponds to the material's deformation capacity, while EM evaluates the film rigidity [44,55]. Improved mechanical properties, structural integrity, and better flexibility are expected in polymer-based composite materials [56].

The results of the mechanical parameters evaluated (TS, %EAB and EM) are shown in Table 2. In the films where the natural extract was not incorporated (chitosan and chitosan + MMT) the nanoclay increased the tensile strength of the biopolymers as well as their plasticity ($p < 0.05$), without, however, interfering with the stiffness of the samples ($p > 0.05$; Table 2). This can be attributed to the uniform dispersion of MMT in the chitosan matrix (i.e., achievement of an exfoliated or intercalated conformation) and the strong interaction between the polymer and montmorillonite, as previously discussed.

The incorporation of ginger essential oil resulted in films that were less resistant to traction, more elastic, and less rigid. Regarding TS, there was a reduction between 0.6% and 31.6% (for films containing between 0.5% and 2% GEO; Table 2). The incorporation of MMT did not result in difference in the TS of the films also incorporated with GEO ($p > 0.05$). The incorporation of lipids into the polymeric matrix of films induces the formation of heterogeneous and discontinuous structures, which affect the mechanical resistance of the polymers by the partial replacement of the strong polar chemical bonds chitosan–chitosan (between the chitosan molecules) by weaker interactions between chitosan–GEO (active molecules present in GEO) [23].

Table 2. Summary of thickness, mechanical properties, contact angle, water solubility and swelling degree of active bionanocomposites.

GEO (%)	Thickness (μm)		Tensile Strength (MPa)		EAB (%)		Elastic Modulus (GPa)	
	0% MMT	2.5% MMT	0% MMT	2.5% MMT	0% MMT	2.5% MMT	0% MMT	2.5% MMT
0	41.7 ± 2.9 [Da*]	39.5 ± 2.0 [Ca]	46.7 ± 1.7 [Ab]	66.6 ± 3.1 [Aa]	17.9 ± 1.2 [Bb]	33.5 ± 2.5 [Aa]	2.05 ± 0.19 [Aa]	1.86 ± 0.14 [Aa]
0.5	55.5 ± 1.3 [Ca]	52.5 ± 0.6 [Bb]	46.3 ± 5.1 [Aa]	42.0 ± 4.2 [Ba]	23.8 ± 5.3 [ABa]	22.3 ± 2.9 [Ba]	$1.66 \pm .012$ [Aa]	1.30 ± 0.30 [Ba]
1.0	68.2 ± 3.1 [Ba]	67.4 ± 1.1 [Aa]	27.2 ± 5.8 [Ba]	34.6 ± 1.0 [BCa]	36.0 ± 9.4 [Aa]	33.1 ± 2.3 [Aa]	0.30 ± 0.09 [Cb]	0.88 ± 0.13 [BCa]
2.0	81.3 ± 4.5 [Aa]	68.6 ± 0.9 [Ab]	32.1 ± 3.9 [Ba]	30.3 ± 0.7 [Ca]	33.6 ± 6.6 [ABa]	35.8 ± 1.3 [Aa]	0.71 ± 0.15 [Ba]	0.47 ± 0.05 [Ca]

GEO (%)	Water Solubility (%)		Swelling Degree (%)		Contact Angle (Degrees)	
	0% MMT	2.5% MMT	0% MMT	2.5% MMT	0% MMT	2.5% MMT
0	23.1 ± 1.6 [Aa]	20.4 ± 0.6 [Bb]	132.3 ± 10.0 [Ab*]	192.2 ± 13.3 [Aa]	70.9 ± 4.0 [Ab]	90.6 ± 5.5 [Aa]
0.5	20.3 ± 0.4 [Ab]	21.9 ± 0.6 [Ba]	124.6 ± 3.5 [Aa]	35.1 ± 20.7 [Cb]	70.3 ± 7.5 [Aa]	68.6 ± 1.6 [Ba]
1.0	16.4 ± 0.4 [Bb]	18.7 ± 0.2 [Ba]	97.8 ± 18.2 [Ba]	97.0 ± 13.4 [Ba]	66.6 ± 5.5 [Aa]	70.9 ± 3.2 [Ba]
2.0	19.6 ± 2.1 [ABa]	23.5 ± 2.4 [Aa]	87.0 ± 2.5 [Ba]	54.7 ± 20.5 [BCa]	57.2 ± 1.9 [Ba]	63.8 ± 7.6 [Ba]

(A–D): Within each parameter, values in the same column not sharing upper case superscript letters indicate statistically significant differences among formulations ($p < 0.05$). (a–b): Within each parameter, values in the same line not sharing lower case superscript letters indicate statistically significant differences among formulations ($p < 0.05$). Ginger Essential Oil (GEO); Sodium Montmorillonite (MMT); Elongation at break (EAB).

Several authors reported a reduction in TS of different biopolymers due to the incorporation of essential oils into their structure (ginger and cinnamon EOs in sodium caseinate films [57]; oregano EO in triticale protein films [58]; thyme, rosemary, and oregano EOs in polylactic acid (PLA) films [59], or essential oils extracted from roots (gingers and saffron) incorporated into gelatin films extracted from fish skin [47]). This behavior was also observed, however with less intensity in chitosan films incorporated with REO [16], probably due to the different chemical compositions of the oils [23].

Regarding the elongation at break, the incorporation of GEO resulted in films with increased extensibility ($p < 0.05$), probably due to the discontinuity of the polymer matrix due to the weaker interactions between chitosan molecules and essential oil components, such as discussed above. The incorporation of MMT, again, did not result in changes in this property ($p > 0.05$), as the %EAB of the films with the same concentrations of GEO, incorporated with MMT or not, did not differ from each other (Table 2).

Therefore, the ginger essential oil probably acted as a plasticizer, as it decreased the strength of the films while increasing their plasticity (ability to stretch before tearing/breaking). In addition, this plasticizing effect of EOs in biopolymers is influenced by the contents of the extracts incorporated in the polymer matrix [47]; as observed in the results (Table 2), greater changes in the mechanical properties were observed with the higher GEO concentrations incorporated.

Materials with lower tensile strength and higher plasticity (%EAB) are also less rigid, which was also observed for chitosan films with the incorporation of GEO (statistically reduced the elastic modulus, $p < 0.05$; Table 2). Similar behavior was also observed in chitosan films incorporated with cinnamon essential oil [52] or in PLA films incorporated with different types of essential oils [59].

According to Zeid et. al. (2019) [59], it is difficult to compare results with different types of materials and additives under different processing conditions; thus, it is crucial to have more available data regarding these novel materials. Furthermore, the application of other techniques to study the rheology of the film form dispersion may contribute with valuable information for understanding the materials' behavior, as it will influence their final specific application [56].

3.6. Optical Properties

It is important to study the optical properties of food contact material because these may interfere on consumer acceptance of the food packaged, as the film can affect the general appearance of the product inside the packaging [53].

Optical properties are shown in Table 3. The incorporation of GEO or MMT decreased the hue value (i.e., films with a more yellow color) while increasing film opacity and color saturation (i.e., greater chromaticity, $p < 0.05$). Similar results were observed in bionanocomposites of chitosan/montmorillonite incorporated with REO [16], as well as those reported for chitosan films incorporated with cinnamon EO [52], citronella, or thyme EOs [53].

The chemical bonding of different molecules to the polymer chain also modifies the properties of the material in terms of light absorption [60], as observed in chitosan films incorporated with natural extracts rich in phenolic compounds [61]. According to Acevedo-Fani et al. (2015) [44], the oil droplets present internally in the polymer matrix can increase the light scattering at the interface of the droplets, resulting in an increase in the opacity values of the material.

Figure 4 depicts the scanning spectra of the percentage of light transmitted through the film between wavelengths of 190–900 nm. Both the incorporation of GEO and MMT reduced the transparency of the films. A high barrier to UV light (wavelengths less than 350 nm) is desirable in food packaging materials since these can be applied to the preservation of oxidative processes [17]. However, in the wavelengths of visible light, the greater the opacity of the material, the worse is its acceptance by the consumer (who always looks for transparent films capable of exposing the packaged product) [53]. Thus, the bionanocomposites produced have this disadvantage from the visual point of view, despite the positive aspect of greater protection of the packaged food (against light, as a secondary antioxidant material).

Table 3. Summary of barrier and optical properties of the active bionanocomposite.

Film	WVP (10^{-11} mol/m·s·Pa)	OP (10^{-16} mol/m·s·Pa)	Cromaticity	Hue *	Opacity
Ch	1.40 ± 0.09 [C*]	0.184 ± 0.052 [DE]	3.1 ± 0.1 [D]	129.0 ± 1.0 [A]	1.1 ± 0.2 [C]
Ch + MMT	1.75 ± 0.10 [BC]	0.098 ± 0.008 [F]	4.2 ± 0.4 [C]	118.8 ± 1.8 [B]	1.7 ± 0.3 [C]
Ch + 0.5% GEO	1.93 ± 0.36 [ABC]	0.182 ± 0.008 [DE]	5.8 ± 0.3 [B]	112.3 ± 0.6 [C]	2.6 ± 0.1 [B]
Ch + MMT + 0.5% GEO	1.94 ± 0.27 [ABC]	0.171 ± 0.001 [E]	6.8 ± 0.6 [B]	109.8 ± 1.4 [CD]	3.1 ± 0.7 [B]
Ch + 1% GEO	1.95 ± 0.21 [ABC]	0.255 ± 0.010 [BC]	7.0 ± 0.8 [B]	110.2 ± 1.3 [CD]	4.1 ± 0.1 [AB]
Ch + MMT + 1% GEO	2.12 ± 0.08 [AB]	0.246 ± 0.013 [CD]	8.0 ± 0.2 [AB]	107.9 ± 0.1 [DE]	5.0 ± 0.2 [AB]
Ch + 2% GEO	1.94 ± 0.10 [ABC]	0.325 ± 0.037 [A]	9.6 ± 1.9 [A]	107.1 ± 1.8 [DE]	4.2 ± 1.4 [AB]
Ch + MMT+ 2% GEO	2.41 ± 0.16 [A]	0.285 ± 0.015 [AB]	9.7 ± 1.3 [A]	105.1 ± 0.7 [E]	6.2 ± 1.8 [A]

(A–E): values in the same column not sharing upper case superscript letters indicate statistically significant differences among formulations ($p < 0.05$). Chitosan (Ch); ginger essential oil (OEG); sodium montmorillonite (MMT); oxygen permeability (OP); water vapor permeability (WVP).

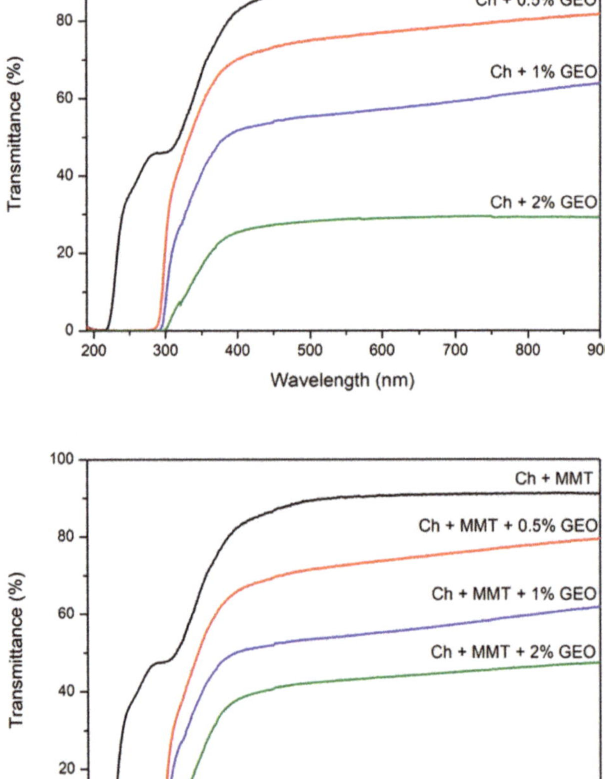

Figure 4. Film transparency of biopolymers produced. Ch, chitosan; MMT, sodium montmorillonite; GEO, ginger essential oil.

3.7. Solubility in Water, Swelling Degree and Contact Angle

The study of the film's solubility in water, contact angle, and swelling degree with water plays an important role when developing novel materials for food applications, as it provides insights on material behavior in contact with aqueous food matrices and understanding of its resistance under this conditions [16,62].

The water solubility of bionanocomposites was affected by both factors evaluated (the incorporation of GEO and MMT; Table 2). Without the incorporation of MMT, only the films added with 1% of GEO presented lower values of solubility in water ($p < 0.05$) in comparison with the control film, probably due to phenolic compounds (present in GEO) cross-linking in the chitosan chain [27,63] that reduced the release of polymer toward the water. Similar results were reported in the literature for biopolymers incorporated with natural compounds [51,64]. When GEO and MMT were incorporated into the films, only for samples with higher GEO content showed a significant increase in this parameter ($p < 0.05$). In general, MMT when associated with GEO contributed to an increase in the water solubility of the films, whereas in the films with GEO but without addition of the nanoclay, a tendency was observed for the decrease in this parameter.

Regarding the films' ability to absorb water, the incorporation of GEO resulted in a decrease in the swelling index ($p < 0.05$). The addition of montmorillonite enhanced even more the effect of diminishing the films' water-absorbing capacity. The formation of crosslinks between the active compounds of GEO and chitosan, as well as the good interaction of sodium montmorillonite with chitosan (due to the intercalated conformation achieved) and GEO, are responsible for the decrease in the ability of the films to absorb water. The chemical interactions created in the polymer matrix block the reactive groups of the chitosan from reacting with the water, thus diminishing its capacity to absorb it [27,36,39,64].

Analyzing the contact angle of the films without the nanofiller, a tendency to increase film surface hydrophilicity (reduction of CA) with the incorporation of GEO was observed; however this difference was only statistically significant for the films with the highest concentration of GEO (2%, $p < 0.05$; Table 2). For the films incorporated with both GEO and MMT, a decrease in the contact angle was also observed; however, in this case, the differences were statistically significant when compared to chitosan + MMT ($p < 0.05$), but not within different levels of GEO ($p > 0.05$). These results indicate the presence of hydrophilic substances on the surface of the films, probably phenolic compounds of GEO that enhanced the interaction with the water droplets [65].

3.8. Barrier Properties

It is desirable that packaging materials exhibit good barrier properties (against light, gases, or water vapor) in order to perform their function as outer physical protection against the external environment, enabling the shelf life extension of the food products packaged [3,66].

Chitosan, a cationic polysaccharide, is a polymer with strong interactions in the polymer chain, which often restrict/decrease its movement, resulting in good oxygen barriers [67]. However, hydrogen bonds with water are also likely to happen, and water absorption occurs (water absorption breaks the intermolecular interactions between the polymer chains), so under conditions of high relative humidity the transmission rates increase, and consequently the permeability as well [68].

The film's water vapor permeability did not vary with the incorporation of MMT or GEO ($p > 0.05$), except for the film chitosan + 2% GEO + MMT that presented a lower barrier to water vapor when compared to chitosan or chitosan + MMT films ($p < 0.05$); however, this difference was not significant in relation to the other films also incorporated with GEO or GEO + MMT ($p > 0.05$; Table 3). Similar results were reported in a recent study with PLA incorporated with different essential oils at 10% (*w/w*) [59], the authors used thyme, rosemary, or oregano EOs to add antioxidant properties to the bio-based films and did not found statistical differences in the composites' WPV.

Despite the non-statistical differences in the WPV, a tendency of increase in permeability by the incorporation of both GEO and MMT was observed (Table 3). However, the effect of the GEO concentration on this change was not observed when MMT was not incorporated (WPV of films without MMT but with different levels of GEO were practically the same, around 1.94×10^{-11} mol /m·s·Pa).

Contrary to what was observed, the incorporation of hydrophobic substances (such as oils) in the films should reduce the WPV, since the water vapor transfer process depends on the ratio between hydrophilic/hydrophobic constituents [69]. However, similar results were found by Atarés et al. (2011) [68], who observed an increase in WPV of hydroxymethyl cellulose films incorporated with GEO when the tests were performed at 35 °C, a similar temperature to the one used in our assay. According to these authors, the physical state in which the essential oil is found is a determinant in the effects caused in the WPV. For low temperatures, a decrease in permeability occurs, whereas for higher temperatures, when the oil is in the liquid state, it can favor the molecular mobility of the polymer chain, promoting the transport of molecules through the emulsified film [68]. Perdones et al. (2014) [52] attributed the increase in WPV of chitosan films incorporated with cinnamon essential oil to the possible interactions between the EO and chitosan components that made the polymer matrix more open to the transport of water molecules, and, at the same time, they plasticize the film. These results corroborate with the present WPV observed for the bionacomposites produced.

Regarding oxygen permeability, the incorporation of GEO reduced this gas barrier (Table 3). Similar results were observed when rosemary essential oil was incorporated in chitosan/montmorillonite composites [16]. It is possible that GEO acted as a plasticizer to the chitosan films, as it increased the elongation capacity of the material (%EAB, as previously discussed), thus increasing both the permeability to oxygen and to water vapor. The plasticizers act as an internal lubricant, reducing the frictional forces between the polymer chains, and increasing the intermolecular space, thus allowing a greater mobility of the polymer chains and consequently facilitating the transport of gases [52,70].

A similar behavior was observed in chitosan films incorporated with cinnamon essential oil [52], and in films of hydroxypropylmethyl cellulose incorporated with GEO [68]. The liquid state of the essential oils as well as their hydrophobic character facilitate the transport of the oxygen through the film due to the increase of its solubility in the polymer matrix [68]. Montmorillonite helped to reduce the negative effect of incorporating GEO into the oxygen barrier. However, compared with commercial EVOH film, which is considered to be one of the best oxygen barrier packages, films incorporated with GEO exhibit OP in the same order of magnitude as EVOH (0.24×10^{-16} mol/m·s·Pa) [71], demonstrating the potential of these films to protect against exposure to oxygen, and consequently to oxidative processes catalyzed by this gas. These results are in line with the results obtained from fresh poultry in an in situ study [28], in which samples wrapped with these bionanocomposites maintained their color and pH values and the thiobarbituric acid reactive substance index (TBARS) increased at a lower rate, helping to extend poultry meat shelf-life.

4. Conclusions

Homogeneous, transparent, yellowish, thin bionanocomposite films were successfully produced by casting. Furthermore, the homogenization process used to produce the films has proven to be adequate in achieving a good exfoliation of the MMT into the polymer chains. The good interaction between the composite's components (chitosan, MMT, and GEO constitutes) was demonstrated by the microscopy images, XRD, and FTIR results. The incorporation of GEO resulted in some irregularities on the film surface and changed the internal structure into a more "sponge like" shape. Overall, the incorporation of GEO resulted in films less resistant and more plastic, with lower barrier properties. However, the incorporation of MMT counterbalanced this effect. Thus, this work also shows that the incorporation of EOs into food packaging polymeric matrices is an interesting approach, as it reduces the EO amount necessary for preserving the foodstuff, therefore reducing costs and overcoming the aroma problem generally related to the directly incorporation of EOs into food. These novel bionanocomposites, therefore, have potential to be used as active packaging to preserve food products, but their application should also take into consideration their physical properties.

Coatings **2019**, *9*, 700

Author Contributions: Conceptualization, V.G.L.S., M.P.D., I.M.C. and A.L.F.; Data curation, V.G.L.S.; Formal analysis, V.G.L.S., J.R.A.P., C.R., P.F.R., A.L., R.J.S. and J.C.; Investigation, V.G.L.S.; Methodology, V.G.L.S., J.R.A.P., C.R., P.F.R., A.L., R.J.S., J.C., M.P.D., I.M.C. and A.L.F.; Supervision, M.P.D., F.B.F., I.M.C. and A.L.F.; Writing–original draft, V.G.L.S., J.R.A.P., C.R., P.F.R., A.L., R.J.S., J.C., M.P.D., F.B.F., I.M.C. and A.L.F.; Writing–review and editing, M.P.D., F.B.F., I.M.C. and A.L.F.

Funding: This work was supported by CNPq – Brazil (grant number 200790/2014-5) and the MEtRICs unit which is financed by national funds from FCT/MCTES (UID/EMS/04077/2019). This work was also supported by the Associate Laboratory for Green Chemistry- LAQV which is financed by national funds from FCT/MCTES (UID/QUI/50006/2019) and UCIBIO, which is funded by national funds from FCT/MCTES (UID/Multi/04378/2019). It is also acknowledged the funding of CENIMAT by FEDER through the program COMPETE 2020 and National Funds through FCT-Portuguese Foundation for Science and Technology, under the project UID/CTM/50025/2019).

Acknowledgments: The authors acknowledge BYK Additives & Instruments for donating the sodium montmorillonite (Cloisite®Na+).

Conflicts of Interest: The authors declare no conflict of interest. The founding sponsors had no role in the design of the study; in the collection, analyses, or interpretation of data; in the writing of the manuscript, and in the decision to publish the results.

References

1. Hassan, B.; Chatha, S.A.S.; Hussain, A.I.; Zia, K.M.; Akhtar, N. Recent advances on polysaccharides, lipids and protein based edible films and coatings: A review. *Int. J. Biol. Macromol.* **2018**, *109*, 1095–1107. [CrossRef] [PubMed]

2. Azeredo, H.M.C. De Nanocomposites for food packaging applications. *Food Res. Int.* **2009**, *42*, 1240–1253. [CrossRef]

3. Vilarinho, F.; Andrade, M.; Buonocore, G.G.; Stanzione, M.; Vaz, M.F.; Sanches Silva, A. Monitoring lipid oxidation in a processed meat product packaged with nanocomposite poly(lactic acid) film. *Eur. Polym. J.* **2018**, *98*, 362–367. [CrossRef]

4. Fernando, A.L.; Duarte, M.P.; Vatsanidou, A.; Alexopoulou, E. Environmental aspects of fiber crops cultivation and use. *Ind. Crops Prod.* **2015**, *68*, 105–115. [CrossRef]

5. Ferreira, A.R.V.; Torres, C.A.V.; Freitas, F.; Sevrin, C.; Grandfils, C.; Reis, M.A.M.; Alves, V.D.; Coelhoso, I.M. Development and characterization of bilayer films of FucoPol and chitosan. *Carbohydr. Polym.* **2016**, *147*, 8–15. [CrossRef] [PubMed]

6. Souza, V.G.L.; Pires, J.R.A.; Vieira, É.T.; Coelhoso, I.M.; Duarte, M.P.; Fernando, A.L. Activity of chitosan-montmorillonite bionanocomposites incorporated with rosemary essential oil: From in vitro assays to application in fresh poultry meat. *Food Hydrocoll.* **2019**, *89*, 241–252. [CrossRef]

7. Souza, V.G.L.; Fernando, A.L. Nanoparticles in food packaging: Biodegradability and potential migration to food—A review. *Food Packag. Shelf Life* **2016**, *8*, 63–70. [CrossRef]

8. Pires, J.R.A.; de Souza, V.G.L.; Fernando, A.L. Chitosan/montmorillonite bionanocomposites incorporated with rosemary and ginger essential oil as packaging for fresh poultry meat. *Food Packag. Shelf Life* **2018**, *17*, 142–149. [CrossRef]

9. Siripatrawan, U.; Noipha, S. Active film from chitosan incorporating green tea extract for shelf life extension of pork sausages. *Food Hydrocoll.* **2012**, *27*, 102–108. [CrossRef]

10. Dutta, P.K.; Tripathi, S.; Mehrotra, G.K.; Dutta, J. Perspectives for chitosan based antimicrobial films in food applications. *Food Chem.* **2009**, *114*, 1173–1182. [CrossRef]

11. Darder, M.; Colilla, M.; Ruiz-Hitzky, E. Biopolymer−Clay Nanocomposites Based on Chitosan Intercalated in Montmorillonite. *Chem. Mater.* **2003**, *15*, 3774–3780. [CrossRef]

12. Sánchez-González, L.; Cháfer, M.; Hernández, M.; Chiralt, A.; González-Martínez, C. Antimicrobial activity of polysaccharide films containing essential oils. *Food Control* **2011**, *22*, 1302–1310. [CrossRef]

13. Chivrac, F.; Pollet, E.; Avérous, L. Progress in nano-biocomposites based on polysaccharides and nanoclays. *Mater. Sci. Eng. R Reports* **2009**, *67*, 1–17. [CrossRef]

14. Alizadeh, V.; Barzegar, H.; Nasehi, B.; Samavati, V. Development of a chitosan-montmorillonite nanocomposite film containing Satureja hortensis essential oil. *Iran. Food Sci. Technol. Res. J.* **2018**, *13*, 131–143. [CrossRef]

15. Abdollahi, M.; Rezaei, M.; Farzi, G. Influence of chitosan/clay functional bionanocomposite activated with rosemary essential oil on the shelf life of fresh silver carp. *Int. J. Food Sci. Technol.* **2014**, *49*, 811–818. [CrossRef]

16. Souza, V.G.L.; Pires, J.R.A.; Rodrigues, P.F.; Lopes, A.A.S.; Fernandes, F.M.B.; Duarte, M.P.; Coelhoso, I.M.; Fernando, A.L. Bionanocomposites of chitosan/montmorillonite incorporated with Rosmarinus officinalis essential oil: Development and physical characterization. *Food Packag. Shelf Life* **2018**, *16*, 148–156. [CrossRef]

17. Souza, V.G.L.; Rodrigues, P.F.; Duarte, M.P.; Fernando, A.L. Antioxidant Migration Studies in Chitosan Films Incorporated with Plant Extracts. *J. Renew. Mater.* **2018**, *6*, 548–558. [CrossRef]

18. Regnier, T.; Combrinck, S.; Du Plooy, W. Essential Oils and Other Plant Extracts as Food Preservatives. In *Progress in Food Preservation*; Wiley-Blackwell: Oxford, UK, 2012; pp. 539–579.

19. Pascoal, A.; Quirantes-Piné, R.; Fernando, A.L.; Alexopoulou, E.; Segura-Carretero, A. Phenolic composition and antioxidant activity of kenaf leaves. *Ind. Crops Prod.* **2015**, *78*, 116–123. [CrossRef]

20. Souza, V.G.L.; Rodrigues, C.; Ferreira, L.; Pires, J.R.A.; Duarte, M.P.; Coelhoso, I.; Fernando, A.L. In vitro bioactivity of novel chitosan bionanocomposites incorporated with different essential oils. *Ind. Crops Prod.* **2019**, *140*, 111563. [CrossRef]

21. Appendini, P.; Hotchkiss, J.H. Review of antimicrobial food packaging. *Innov. Food Sci. Emerg. Technol.* **2002**, *3*, 113–126. [CrossRef]

22. Pola, C.C.; Medeiros, E.A.A.; Pereira, O.L.; Souza, V.G.L.; Otoni, C.G.; Camilloto, G.P.; Soares, N.F.F. Cellulose acetate active films incorporated with oregano (*Origanum vulgare*) essential oil and organophilic montmorillonite clay control the growth of phytopathogenic fungi. *Food Packag. Shelf Life* **2016**, *9*, 69–78. [CrossRef]

23. Atarés, L.; Chiralt, A. Essential oils as additives in biodegradable films and coatings for active food packaging. *Trends Food Sci. Technol.* **2016**, *48*, 51–62. [CrossRef]

24. Ribeiro-Santos, R.; Andrade, M.; de Melo, N.R.; Sanches-Silva, A. Use of essential oils in active food packaging: Recent advances and future trends. *Trends Food Sci. Technol.* **2017**, *61*, 132–140. [CrossRef]

25. Noori, S.; Zeynali, F.; Almasi, H. Antimicrobial and antioxidant efficiency of nanoemulsion-based edible coating containing ginger (*Zingiber officinale*) essential oil and its effect on safety and quality attributes of chicken breast fillets. *Food Control* **2018**, *84*, 312–320. [CrossRef]

26. Silva, F.; Domingues, F.C.; Nerín, C. Trends in microbial control techniques for poultry products. *Crit. Rev. Food Sci. Nutr.* **2016**, *58*, 1–19. [CrossRef]

27. Souza, V.G.L.; Fernando, A.L.; Pires, J.R.A.; Rodrigues, P.F.; Lopes, A.A.S.; Fernandes, F.M.B. Physical properties of chitosan films incorporated with natural antioxidants. *Ind. Crops Prod.* **2017**, *107*, 565–572. [CrossRef]

28. Souza, V.G.L.; Pires, J.R.; Vieira, É.T.; Coelhoso, I.M.; Duarte, M.P.; Fernando, A.L. Shelf Life Assessment of Fresh Poultry Meat Packaged in Novel Bionanocomposite of Chitosan/Montmorillonite Incorporated with Ginger Essential Oil. *Coatings* **2018**, *8*, 177. [CrossRef]

29. Woranuch, S.; Yoksan, R. Eugenol-loaded chitosan nanoparticles: II. Application in bio-based plastics for active packaging. *Carbohydr. Polym.* **2013**, *96*, 586–592. [CrossRef]

30. Higueras, L.; López-carballo, G.; Cerisuelo, J.P.; Gavara, R. Preparation and characterization of chitosan/HP-β-cyclodextrins composites with high sorption capacity for carvacrol. *Carbohydr. Polym.* **2012**, *97*, 262–268. [CrossRef]

31. ASTM—America Society Standard Testing and Materials. *Standard Test Method for Tensile Properties of Thin Plastic Sheeting—D882-12*; ASTM: West Conshohocken, PA, USA, 2012; p. 12.

32. Pastor, C.; Sánchez-González, L.; Chiralt, A.; Cháfer, M.; González-Martínez, C. Physical and antioxidant properties of chitosan and methylcellulose based films containing resveratrol. *Food Hydrocoll.* **2013**, *30*, 272–280. [CrossRef]

33. Park, S.-I.; Zhao, Y. Incorporation of a high concentration of mineral or vitamin into chitosan-based films. *J. Agric. Food Chem.* **2004**, *52*, 1933–1939. [CrossRef] [PubMed]

34. Kanatt, S.R.; Rao, M.S.; Chawla, S.P.; Sharma, A. Active chitosan–polyvinyl alcohol films with natural extracts. *Food Hydrocoll.* **2012**, *29*, 290–297. [CrossRef]

35. Alves, V.D.; Costa, N.; Coelhoso, I.M. Barrier properties of biodegradable composite films based on kappa-carrageenan/pectin blends and mica flakes. *Carbohydr. Polym.* **2010**, *79*, 269–276. [CrossRef]

36. Lavorgna, M.; Piscitelli, F.; Mangiacapra, P.; Buonocore, G.G. Study of the combined effect of both clay and glycerol plasticizer on the properties of chitosan films. *Carbohydr. Polym.* **2010**, *82*, 291–298. [CrossRef]

37. Dias, M.V.; Machado Azevedo, V.; Borges, S.V.; Soares, N.D.F.F.; de Barros Fernandes, R.V.; Marques, J.J.; Medeiros, É.A.A. Development of chitosan/montmorillonite nanocomposites with encapsulated α-tocopherol. *Food Chem.* **2014**, *165*, 323–329. [CrossRef]

38. Wang, S.F.; Shen, L.; Tong, Y.J.; Chen, L.; Phang, I.Y.; Lim, P.Q.; Liu, T.X. Biopolymer chitosan/montmorillonite nanocomposites: Preparation and characterization. *Polym. Degrad. Stab.* **2005**, *90*, 123–131. [CrossRef]

39. Silva-Weiss, A.; Bifani, V.; Ihl, M.; Sobral, P.J.A.; Gómez-Guillén, M.C. Structural properties of films and rheology of film-forming solutions based on chitosan and chitosan-starch blend enriched with murta leaf extract. *Food Hydrocoll.* **2013**, *31*, 458–466. [CrossRef]

40. Ávila, A.; Bierbrauer, K.; Pucci, G.; López-González, M.; Strumia, M. Study of optimization of the synthesis and properties of biocomposite films based on grafted chitosan. *J. Food Eng.* **2012**, *109*, 752–761. [CrossRef]

41. Bonilla, J.; Sobral, P.J.A. Investigation of the physicochemical, antimicrobial and antioxidant properties of gelatin-chitosan edible film mixed with plant ethanolic extracts. *Food Biosci.* **2016**, *16*, 17–25. [CrossRef]

42. Qin, Y.Y.; Zhang, Z.H.; Li, L.; Yuan, M.L.; Fan, J.; Zhao, T.R. Physio-mechanical properties of an active chitosan film incorporated with montmorillonite and natural antioxidants extracted from pomegranate rind. *J. Food Sci. Technol.* **2015**, *52*, 1471–1479. [CrossRef]

43. Sun, L.; Sun, J.; Chen, L.; Niu, P.; Yang, X.; Guo, Y. Preparation and characterization of chitosan film incorporated with thinned young apple polyphenols as an active packaging material. *Carbohydr. Polym.* **2017**, *163*, 81–91. [CrossRef] [PubMed]

44. Acevedo-Fani, A.; Salvia-Trujillo, L.; Rojas-Graü, M.A.; Martín-Belloso, O. Edible films from essential-oil-loaded nanoemulsions: Physicochemical characterization and antimicrobial properties. *Food Hydrocoll.* **2015**, *47*, 168–177. [CrossRef]

45. Sakurai, K.; Maegawa, T.; Takahashi, T. Glass transition temperature of chitosan and miscibility of chitosan/poly(N-vinyl pyrrolidone) blends. *Polymer* **2000**, *41*, 7051–7056. [CrossRef]

46. Woranuch, S.; Yoksan, R. Eugenol-loaded chitosan nanoparticles: I. Thermal stability improvement of eugenol through encapsulation. *Carbohydr. Polym.* **2013**, *96*, 578–585. [CrossRef] [PubMed]

47. Tongnuanchan, P.; Benjakul, S.; Prodpran, T. Physico-chemical properties, morphology and antioxidant activity of film from fish skin gelatin incorporated with root essential oils. *J. Food Eng.* **2013**, *117*, 350–360. [CrossRef]

48. Noshirvani, N.; Ghanbarzadeh, B.; Gardrat, C.; Rezaei, M.R.; Hashemi, M.; Le Coz, C.; Coma, V. Cinnamon and ginger essential oils to improve antifungal, physical and mechanical properties of chitosan-carboxymethyl cellulose films. *Food Hydrocoll.* **2017**, *70*, 36–45. [CrossRef]

49. Lewandowska, K.; Sionkowska, A.; Kaczmarek, B.; Furtos, G. Characterization of chitosan composites with various clays. *Int. J. Biol. Macromol.* **2014**, *65*, 534–541. [CrossRef]

50. Rimdusit, S.; Jingjid, S.; Damrongsakkul, S.; Tiptipakorn, S.; Takeichi, T. Biodegradability and property characterizations of Methyl Cellulose: Effect of nanocompositing and chemical crosslinking. *Carbohydr. Polym.* **2008**, *72*, 444–455. [CrossRef]

51. Beigzadeh Ghelejlu, S.; Esmaiili, M.; Almasi, H. Characterization of chitosan-nanoclay bionanocomposite active films containing milk thistle extract. *Int. J. Biol. Macromol.* **2016**, *86*, 613–621. [CrossRef]

52. Perdones, Á.; Vargas, M.; Atarés, L.; Chiralt, A. Physical, antioxidant and antimicrobial properties of chitosan–cinnamon leaf oil films as affected by oleic acid. *Food Hydrocoll.* **2014**, *36*, 256–264. [CrossRef]

53. Peng, Y.; Li, Y. Combined effects of two kinds of essential oils on physical, mechanical and structural properties of chitosan films. *Food Hydrocoll.* **2014**, *36*, 287–293. [CrossRef]

54. Wang, Y.; Xia, Y.; Zhang, P.; Ye, L.; Wu, L.; He, S. Physical Characterization and Pork Packaging Application of Chitosan Films Incorporated with Combined Essential Oils of Cinnamon and Ginger. *Food Bioprocess Technol.* **2017**, *10*, 503–511. [CrossRef]

55. Sarantópoulos, C.G.L.; Oliveira, L.M.; Padula, M.; Coltro, L.; Aalves, R.M.V.; Garcia, E.C. *Embalagens Plásticas Flexíveis: Principais Polímeros e Avaliação de Propriedades*; CETEA/ITAL: Campinas, Brazil, 2002; p. 267.

56. Russo, T.; D'Amora, U.; Gloria, A.; Tunesi, M.; Sandri, M.; Rodilossi, S.; Albani, D.; Forloni, G.; Giordano, C.; Cigada, A.; et al. Systematic analysis of injectable materials and 3D rapid prototyped magnetic scaffolds: From CNS applications to soft and hard tissue repair/regeneration. *Procedia Eng.* **2013**, *59*, 233–239. [CrossRef]

57. Atarés, L.; Bonilla, J.; Chiralt, A. Characterization of sodium caseinate-based edible films incorporated with cinnamon or ginger essential oils. *J. Food Eng.* **2010**, *100*, 678–687. [CrossRef]

58. Aguirre, A.; Borneo, R.; León, A.E. Antimicrobial, mechanical and barrier properties of triticale protein films incorporated with oregano essential oil. *Food Biosci.* **2013**, *1*, 2–9. [CrossRef]

59. Zeid, A.; Karabagias, I.K.; Nassif, M.; Kontominas, M.G. Preparation and evaluation of antioxidant packaging films made of polylactic acid containing thyme, rosemary, and oregano essential oils. *J. Food Process. Preserv.* **2019**, 1–11. [CrossRef]

60. Moradi, M.; Tajik, H.; Razavi Rohani, S.M.; Oromiehie, A.R.; Malekinejad, H.; Aliakbarlu, J.; Hadian, M. Characterization of antioxidant chitosan film incorporated with Zataria multiflora Boiss essential oil and grape seed extract. *LWT Food Sci. Technol.* **2012**, *46*, 477–484. [CrossRef]

61. Siripatrawan, U.; Harte, B.R. Physical properties and antioxidant activity of an active film from chitosan incorporated with green tea extract. *Food Hydrocoll.* **2010**, *24*, 770–775. [CrossRef]

62. Nunes, C.; Maricato, É.; Cunha, Â.; Nunes, A.; da Silva, J.A.L.; Coimbra, M.A. Chitosan-caffeic acid-genipin films presenting enhanced antioxidant activity and stability in acidic media. *Carbohydr. Polym.* **2013**, *91*, 236–243. [CrossRef]

63. Hafsa, J.; Smach, M.A.; Ben Khedher, M.R.; Charfeddine, B.; Limem, K.; Majdoub, H.; Rouatbi, S. Physical, antioxidant and antimicrobial properties of chitosan films containing Eucalyptus globulus essential oil. *LWT Food Sci. Technol.* **2016**, *68*, 356–364. [CrossRef]

64. Abdollahi, M.; Rezaei, M.; Farzi, G. A novel active bionanocomposite film incorporating rosemary essential oil and nanoclay into chitosan. *J. Food Eng.* **2012**, *111*, 343–350. [CrossRef]

65. Contini, C.; Katsikogianni, M.G.; O'Neill, F.T.; O'Sullivan, M.; Dowling, D.P.; Monahan, F.J. Development of active packaging containing natural antioxidants. *Procedia Food Sci.* **2011**, *1*, 224–228. [CrossRef]

66. Soares, F.; Pires, A.C.S.; Camilloto, G.P.; Santiago-Silva, P.; Espitia, P.J.P.; Silva, W.A. Recent patents on active packaging for food application. *Recent Pat. Food. Nutr. Agric.* **2009**, *1*, 171–178. [CrossRef]

67. Miller, K.S.; Krochta, J.M. Oxygen and aroma barrier properties of edible films: A review. *Trends Food Sci. Technol.* **1997**, *8*, 228–237. [CrossRef]

68. Atarés, L.; Pérez-Masiá, R.; Chiralt, A. The role of some antioxidants in the HPMC film properties and lipid protection in coated toasted almonds. *J. Food Eng.* **2011**, *104*, 649–656. [CrossRef]

69. Baldwin, E.A.; Hagenmaier, R.D.; Bai, J. *Edible Coatings and Films to Improve Food Quality*; CRC Press: Boca Raton, FL, USA, 2012; ISBN 9781138198937.

70. Nouri, A.; Yaraki, M.T.; Ghorbanpour, M.; Agarwal, S.; Gupta, V.K. Enhanced Antibacterial effect of chitosan film using Montmorillonite/CuO nanocomposite. *Int. J. Biol. Macromol.* **2017**, *109*, 1219–1231. [CrossRef]

71. Cerisuelo, J.P.; Alonso, J.; Aucejo, S.; Gavara, R.; Hernández-Muñoz, P. Modifications induced by the addition of a nanoclay in the functional and active properties of an EVOH film containing carvacrol for food packaging. *J. Memb. Sci.* **2012**, *423*, 247–256. [CrossRef]

 *coatings*

Article

Antifungal Starch–Gellan Edible Coatings with Thyme Essential Oil for the Postharvest Preservation of Apple and Persimmon

Mayra Sapper [1],*, Lluís Palou [2], María B. Pérez-Gago [2] and Amparo Chiralt [1]

[1] Instituto Universitario de Ingeniería de Alimentos para el Desarrollo, Universitat Politècnica de València, Camino de Vera s/n, 46022 Valencia, Spain; dchiralt@tal.upv.es

[2] Centro de Tecnología Poscosecha (CTP), Instituto Valenciano de Investigaciones Agrarias (IVIA), 46113 Moncada, Valencia, Spain; palou_llu@gva.es (L.P.); perez_mbe@gva.es (M.B.P.-G.)

* Correspondence: maysap@etsiamn.upv.es; Tel.: +34-644-323-055

Received: 9 May 2019; Accepted: 21 May 2019; Published: 24 May 2019

Abstract: Starch–gellan (80:20) coating formulations were applied to apples and persimmons to analyse their effectiveness controlling the weight loss, respiration rate, fruit firmness, and fungal decay during postharvest. Thyme essential oil (EO) was incorporated (0.25 and 0.5 g per g of polymer) directly or encapsulated in lecithin to enhance antifungal action. Coatings did not reduce weight loss or firmness changes in apples, but they prevented water loss in persimmons. In contrast, no significant effect of the coatings was observed on the respiration rates and the respiration quotient of persimmons, whereas the respiration rates and quotient was increased in apples. On the other hand, the coatings without lecithin reduced the incidence and severity of black spot caused by *Alternaria alternata* in persimmons, regardless of the content of essential oil. Likewise, these reduced the severity of gray mold caused by *Botrytis cinerea* in apples. No positive effect of lecithin in coatings was observed on the postharvest quality and decay in either apples or persimmons, nor did EO exert antifungal action despite its proven effectiveness in in vitro tests.

Keywords: cassava starch; gellan; thyme essential oil; antifungal edible coatings; postharvest; fruit

---

1. Introduction

Postharvest diseases are one of the major factors that affect the quality of horticultural fresh products during storage. Since fruit and vegetables are living organisms, their shelf life is greatly affected by temperature, relative humidity (RH), composition of the atmosphere during and after harvest, and the type and degree of infection by microorganisms or attack by insects [1]. Fruit contains high levels of sugars and nutrient elements, and the low pH values make them particularly susceptible to fungal decay. Fruit fungal infection may occur during flowering, fruit growth, harvesting, transport, packing operations, postharvest storage, or after purchase by the consumer [2]. Moreover, the natural resistance of fruit and vegetables to disease declines with storage duration and ripeness [3].

Gray mold caused by *Botrytis cinerea* Pers. is considered one of the most serious and common postharvest diseases of various fruit, including apples and persimmons [4,5]. The infection may occur in the field, from bloom to harvest, or after harvest, typically causing nests of decay. In fruit that are often stored for extended periods such as apples, field infections that remained latent can resume growth during storage, when the pathogen takes advantage of fruit maturity and environmental conditions and the disease develops (low temperatures and high humidity). In this sense, *B. cinerea* is very well adapted to low temperatures, and it is even able to grow at 0 °C [6]. Infection starts with a darker circular area where the fruit tissues are softer than the other fruit parts, and subsequent abundant sporification, whose colour ranges from white to gray, can develop from the site of infection in conditions of ambient temperature and high humidity [7]. *Alternaria alternata* (Fr.) Keissl. is the causal agent of postharvest black spot in persimmons (*Diospyros kaki* Thunb.) [8], and is generally considered a weak and opportunistic pathogen that gains entry into the fruit via wounds or natural openings, and remains quiescent until the fruit ripens [9]. *A. alternata* and *Penicillium* spp. were found to be the main causal agents of latent and wound infections in persimmons in Spain [5].

After harvest, fresh produce also suffer physiological and biochemical changes that cause detrimental changes in quality and shelf life. Respiration, transpiration, and ethylene production are the main factors contributing to the deterioration of fruits and vegetables [10]. Ethylene is a hormone produced by climacteric fruits, or when fruit undergoes stress, and is partially responsible for changes in the flavour, colour, and texture of fruits and vegetables. In addition, fresh fruits and vegetables lose water during storage due to respiratory and transpiration processes [11]. Water stress also causes metabolic alterations and changes in enzyme activation, causing accelerated senescence, a decline in nutritional value, and increased susceptibility to chilling injury and pathogen invasion. Respiration consists of the oxidative breakdown of organic reserves to simpler molecules, including carbon dioxide (CO_2) and water, with the release of energy [12]. All these biological factors, such as respiration, ethylene production, and resistance to water diffusion depend on the fruit commodity and cultivar, physiological stage at harvest, and storage conditions, which are also related to the composition of the surface waxes. Thus, for example, Morice and Shorland [13] reported that hydrocarbons, alcohols, fatty acids, ursolic acid, and α-farnesene are the main components in natural apple surface waxes, and the amount and composition of these components changed during storage depending on the apple cultivar.

In the last decade, considerable research has been carried out into the development of edible coatings aiming to control the physiological activity of fruit. These coatings can modify the internal gas composition and reduce the water loss through the regulation of oxygen (O_2), CO_2, and water vapour exchange between the fruit and the surrounding atmosphere. However, a certain degree of O_2 and CO_2 permeability is necessary to avoid anaerobic respiration, which induces ethanol production, off-flavour formation, and the loss of produce quality [10]. An additional advantage of edible coatings is the possibility of incorporating food-grade ingredients, such as antimicrobial agents, antioxidants, flavours, colour pigments, and vitamins into the basic formulation with the aim of improving their functional properties.

Traditionally, the postharvest disease control of fresh fruits and vegetables involves the use of synthetic chemical fungicides in those products for which their use is legislated. However, new restrictive regulations regarding fungicide residues, the reduction of the legal acceptability limits of specific fungicides, the emergence of fungicide-resistant strains of the pathogens, and an increasing public concern towards these compounds have led to a global increase in the need to seek safer postharvest alternatives to control the decay of fruits and vegetables [14]. Some of these include antimicrobial antagonists (bacteria, yeast, and fungi) that perform as biocontrol agents, synthetic and natural antimicrobials classified as food-grade additives, or generally recognized as safe (GRAS) compounds, such as organic and inorganic acids and their salts, chitosan, essential oils (EOs), or other plant extracts and different physical methods. Among the natural compounds, EOs and their components have been reported to suppress fungal growth, both in in vitro and in vivo studies. Thus, for example, tea tree, palmarosa and star anise EO vapours completely inhibited the in vitro germination of the apple pathogen *Penicillium expansum* L. [15]; *Melissa officinalis* EO was effective against *B. cinerea*, *P. expansum*, and *Rhizopus stolonifer* (Ehrenb.) Vuill. in in vitro studies [16]. *Pulicaria mauritanica* EO was effective against *Alternaria* sp., *P. expansum*, and *R. stolonifer* [17]. In in vivo studies, the addition of lemon EO enhanced the antifungal activity of chitosan against *B. cinerea* in strawberries [18]; garlic extracts and clove EO treatments reduced the postharvest decay caused by *B. cinerea* and *P. expansum* when applied directly to apples [19]; and a chitosan–oregano EO emulsion exhibited an inhibitory effect on pomegranate fruit inoculated with *Botrytis* sp., but caused some phytotoxicity [20]. Usually, the antibacterial effect of EOs relies on their high content of terpenes and terpenoids and also on the content of other aromatic and aliphatic constituents, all of which are characterized by low molecular weight [21].

However, in spite of the great potential of EOs, the main limitation to their application for decay control is the possible induction of a strong odour or flavour in fruit, phytotoxicity risks, and technological issues associated with commercial-scale fumigations or liquid applications [22]. The addition of EOs to edible coatings based on polymeric matrices could render them more effective at prolonging the postharvest life of horticultural produce, slowing down the diffusion rate of the antimicrobial agent and maintaining a higher concentration of the active compound on the fruit surface for a longer period, while preventing phytotoxicity by avoiding the direct contact of the EO with the fruit skin through the encapsulating action of the polymer. Additionally, fruit coatings can delay or retard the ripening process in climacteric fruit by modifying their internal gas composition and changing their permeability to O_2, CO_2, and ethylene production [23]. Among the different EOs, thyme EO exhibited antifungal action against *B. cinerea* and *A. alternata* when included in starch–gellan films in in vitro studies [24], as well as a complete growth inhibition of *B. cinerea* as vapour in in vitro tests [25]. Gellan, a microbial gum consisting of repeating tetrasaccharide units of glucose, glucuronic acid, and rhamnose residues joined in a linear chain, forms starch–gellan composite films with improved mechanical and barrier properties [24], which could be effective at preserving fruit quality during postharvest storage when applied as coatings. Furthermore, given the antifungal effect of these films when these contained thyme essential oil [24], their application as fruit coatings could represent a good strategy to extend the fruit postharvest life in terms of both quality maintenance and fungal growth inhibition. Nevertheless, in vivo assays in different fruits must be carried out in order to validate the beneficial effect of these coatings.

In this study, starch–gellan coatings incorporating thyme (*Thymus zygis* Loefl. ex L.) EO were applied to apples and persimmons to evaluate: (1) the postharvest behaviour of coated fruit in terms of weight loss, respiration rates, and mechanical properties, and (2) the antifungal efficacy of these coatings applied as a curative treatment against *B. cinerea* in apple and *A. alternata* in persimmon.

2. Materials and Methods

2.1. Reagents

To prepare the coating-forming systems (CFS), cassava starch (S) (with 10% amylose content) (Quimidroga S.A., Barcelona, Spain), low acyl gellan gum (G) (KELCOGEL F, Premium Ingredients, Murcia, Spain), non-GMO soy lecithin with 45% phosphatidylcholine (L) (Lipoid P45, Lipoid GmbH, Ludwigshafen, Germany) and thyme (*T. zygis*) essential oil (Plantis, Artesanía Agrícola SA, Barcelona, Spain) (EO) were used. The glycerol used as plasticizer was supplied by Panreac Química S.A. (Castellar de Vallès, Barcelona, Spain) and the polyoxyethylenesorbitan trioleate (Tween 85®) (T) was purchased from Sigma-Aldrich (Madrid, Spain).

2.2. Preparation of CFS

The formulations were prepared using S and G in a ratio of 8:2, with glycerol as the plasticiser (0.25 g per g of polymer), on the basis of previous studies [26,27]. Firstly, S was dispersed in distilled water and kept at 95 °C for 30 min to induce complete starch gelatinization. Meanwhile, G solution was obtained under stirring at 90 °C for 60 min. Both solutions were cooled down and afterwards, glycerol was added. The S and G systems were mixed to obtain the solutions without EO. The thyme EO (0.25 g per g and 0.5 g per g of polymer), which was used as an antifungal agent, was incorporated, either by direct emulsification or encapsulated in lecithin liposomes (polymer: lecithin ratio of 1:0.5). In the first case, the EO was added directly and the dispersions were homogenized for 3 min at 13.500 rpm using a rotor-stator homogenizer (Ultraturrax Yellow Line DL 25 Basic, IKA, Staufen, Germany). In the second case, the liposome dispersions were previously prepared and added directly to the initial polymer blend solution and kept under soft magnetic stirring for 2 h. A formulation was also obtained with lecithin liposomes without EO, as a control. To obtain the lecithin dispersions, lecithin (5%, w·w^{-1}) was dispersed in distilled water and stirred for at least 4 h at 700 rpm. The EO (2.5% and 5% w·w^{-1}) was added to the lecithin dispersion by using a sonicator (Vibra Cell, Sonics and Materials, Inc., Newtown, CT, USA) at 20 kHz for 10 min with pulses of 1 s, as described by Valencia-Sullca et al. [28]. Tween 85 was also added to S:G CFS (10^5 mg·L^{-1}) in order to ensure the complete wettability of the fruit surface, according to a previous study [29], and tested on apples in a preliminary test. All the solutions were degassed using a vacuum pump (MZ 2C NT, Vacuubrand GmbH + CO KG, Germany). A total of six formulations were obtained: starch:gellan (S:G), control with lecithin (S:G-L), formulations with EO, non-encapsulated (S:G-0.25 and S:G-0.5), and lecithin-encapsulated (S:G-0.25-L and S:G-0.5-L).

2.3. Rheological Behaviour and Contact Angle of the CFS

The rheological behaviour was analysed in triplicate at 25 °C by means of a rotational rheometer (HAAKE Rheostress 1, Thermo Electric Corporation, Karlsruhe, Germany) by using a sensor system of coaxial cylinders, type Z34DIN Ti. Measurements were taken between 0–100 s^{-1}. The obtained data was fitted to the Ostwald de Waale power law model (Equation (1)) in order to determine the consistency (*K*) and the flow behaviour indices (*n*):

$$\sigma = K \cdot \left(\frac{\partial u}{\partial y}\right)^n \tag{1}$$

where σ = shear stress (Pa), K = flow consistency index (Pa·s^n), $\frac{\partial u}{\partial y}$ = shear rate (s^{-1}), and n = the flow behaviour index.

The contact angle (θ) was determined by means of a Dynamic Contact Angle measuring device and Tensiometer (OCA 20, DataPhysics Instruments GmbH, Filderstadt, Germany). For this purpose, thin sections of the skin of the fruit were cut and placed on a glass plate to proceed with the measurements. Then, a droplet of each formulation was placed on the horizontal surface with a needle of 1.19 mm in internal diameter, and the contact angle at the fruit surfaces was measured by the sessile drop method [30]. Measurements were taken in less than 10 s. Image analyses were carried out using SCA20 software. At least 12 replicates were obtained.

2.4. Quality of Coated Fruit

Apples (*Malus domestica* Borkh cv. *Golden Delicious*) and persimmons (*Diospyros kaki* Thunb. cv. *Rojo Brillante*) were purchased from local packinghouses (Valencia, Spain) before any postharvest treatments were applied. Fruit were chosen according to their uniform shape, size, colour, and the absence of surface defects; then, they were subsequently cleaned and disinfected by a 4-min immersion in a 1% sodium hypochlorite solution, thoroughly rinsed with tap water, and air-dried at room temperature before coating application.

CFS were applied manually, using approximately 1.5 mL/fruit, and spread evenly over the fruit surface by using latex glove hands, following the method described by Bai et al. [31]. Water was applied to control fruit to simulate the coating application and its possible effect on the inoculum. Then, each fruit was inspected to assure complete coverage, and all fruit were stored at 25 °C and 65% RH, for 14 days. Ten fruit were considered in each series (coated and non-coated fruits).

2.4.1. Surface Density of Solids (SDS)

The SDS was determined by weighing the samples with a precision balance (Kern PFB 120-3, Germany) before and after coating application to obtain the CFS adhered mass. To calculate the total adhered solids, the mass fraction of each CFS was considered and the SDS ($g \cdot m^{-2}$) was estimated applying Equation (2), according to Marín et al. [32]:

$$SDS = \frac{(m_C - m_0) \cdot X_{sCFS}}{m_0} \cdot \rho \cdot \frac{1}{S_e}$$

(2)

where m_C = mass of the coated apple, m_0 = mass of the uncoated apple, X_{sCFS} = mass fraction of the solids of the CFS (g solids per g of solution), ρ = apple density ($g \cdot cm^{-3}$). To obtain the specific surface ($S_e = 6/d$, m^2 particles per m^3 fruit), the average diameter (d) was calculated considering a spherical geometry for the fruit.

2.4.2. Weight Loss

The weight loss of the fruit during storage was measured using an analytical balance (ME235P, Sartorius, Germany) before and after three, seven, and 14 days of storage. The mass loss was referred to the initial mass of the fruit, and the results were expressed as a relative mass loss rate (day^{-1}), which was obtained from the slope of the fitted straight line to the relative weight loss versus time data. Ten fruits were considered for each formulation and for control fruit.

2.4.3. Respiration Rates

Measurements were taken using a closed system, following the method proposed by Castelló et al. [33], with some modifications. Two apples were placed in hermetic glass jars with a septum in the lid for sampling headspace gas at different times. Gas sampling was carried out every 30 min for 4 h by means of a needle connected to a gas analyser (CheckMate 9900 PBI Dansensor, Ringsted, Denmark). Three replicates per treatment were performed after 7 and 14 days of storage. The respiration rate (R_i) of the samples in terms of CO_2 generation and O_2 consumption was determined from the slope of the fitted linear equation, according to Equation (3). The respiration quotient (RQ) has been determined as the ratio between CO_2 production and the O_2 consumption.

$$y_{it} = y_{i0} \pm 100 \cdot R_i \cdot \frac{M}{V} \cdot t$$

(3)

where y_{it} = gas concentration (%O_2, %CO_2) at time t, y_{i0} = initial gas concentration, R_i = respiration rate ($mL \cdot kg^{-1} \cdot h^{-1}$), M = mass of the samples, V = volume (mL) of headspace, and t = time.

2.4.4. Fruit Firmness

The firmness was measured using a Texture Analyser (Stable Micro Systems, TA.XT plus, Haslemere, England) fitted with an 11-mm diameter probe, applying a modification of the method described by Saei et al. [34]. A small skin area was removed from four opposite sides of each fruit around the equator. The probe penetrated the flesh at 10 mm min^{-1} and the maximum force (F_{max}, N) required to break the flesh was used as fruit firmness. The distance at maximum force (d_{max}, mm) was also taken as another representative parameter of the puncture curve. Ten replicates were used for each formulation after 14 days of storage. The same procedure was applied to uncoated fruit (control), both at the beginning and after 14 days of storage.

2.5. In Vivo Antifungal Assays

For the in vivo assays, *B. cinerea* strain BC03 from the IRTA Culture Collection (Lleida, Catalonia, Spain) was originally isolated from infected grapes from a vineyard located in Lleida and it was deposited at the Spanish Type Culture Collection (CECT-20973) at the University of Valencia (Burjassot, Valencia, Spain). *A. alternata* strain QAV-6 had been isolated from decayed persimmon fruit and maintained in the IVIA CTP Culture Collection of postharvest pathogens (Moncada, Valencia, Spain). These fungal strains were cultured on potato dextrose agar (PDA; Scharlab, Barcelona, Catalonia,

Spain) petri dishes at 25 °C in the dark and used after 7 to 14 days of active growth. Conidia were scraped from the cultures using a sterile loop and subsequently filtered and transferred to test tubes with sterile distilled water and 0.01% Tween 85. The suspensions were adjusted at 1×10^4 conidia mL^{-1} for *B. cinerea* and 5×10^5 conidia mL^{-1} for *A. alternata*, which were selected according to previous experience with these postharvest pathosystems [5,19]. The concentration of conidial suspensions was determined using a haemocytometer.

Fruit were wounded (approximately 1.6 mm in diameter and 2 mm deep) using the tip of a stainless-steel rod once in the fruit equator in the case of apples, and twice in the equator on the same side of the fruit in the case of persimmons (wounds located midway between the calyx and the stem end and 5–6 cm apart). Each wound was inoculated using a micropipette with 20 μL of the correspondent spore suspension 24 h before the application of the coatings (assessment of the coatings' curative activity). As previously described, coatings were applied manually at approximately 1.5 mL per fruit. Air surface drying was allowed at room temperature, and fruit were subsequently placed in perforated plastic trays avoiding direct contact between fruit and incubated at 20 °C and 85 ± 5% RH. Twenty fruit—four replications of five fruit each—were used per treatment. Control fruit were inoculated and treated with water using the same procedure as that for coating application. Lesion diameters (disease severity, mm) were measured after 7 and 12 days of incubation. Disease incidence (%) was expressed as the percentage of infected wounds out of the total number of inoculated wounds per replicate and treatment [6].

2.6. Statistical Analysis

The statistical analyses of the results were performed through an analysis of variance (ANOVA) using Statgraphics Centurion XVI.II (StatPoint Technologies Inc., Warrenton, VA, USA). Fisher's least significant difference (LSD) test was used at the 95% confidence level to determine specific differences between means. Multifactor ANOVA was also used to analyse the effect of the different factors (storage time and type of coating).

3. Results and Discussion

3.1. CFS Properties

The viscosity and contact angle of the different CFS on apple and persimmon skin were analysed since these parameters can affect the coating retention/adhesion on the fruit surface after the coating treatment through their influence on the CFS gravitational drainage before drying and liquid spreadability, all of which affect the coating thickness and homogeneity. The flow curves of the CFS were fitted to the power law model and the rheological parameters (consistency index: K and flow index: n), including the apparent viscosity (η) at 100 s^{-1}, are shown in Table 1. Pseudoplastic behaviour, with similar values of n—lower than 1—was observed in all the cases. Apparent viscosities ranged between 25–42 mPa·s, depending on the CFS composition. Directly-emulsified EO caused an increase in the apparent viscosity of the CFS according to the EO ratio, while lecithin-encapsulated EO reduced the viscosity of the formulations, which is probably due to the smaller droplet size in the encapsulated system [28]. Thus, the S:G-0.5 sample was the most viscous formulation and showed the highest consistency index.

The contact angles of the different CFS on apple and persimmon skin are also shown in Table 1. Values lower than 90° indicate surface wettability, and therefore greater extensibility of the coating on the fruit surface. For a given CFS, the contact angles on the persimmon skin were lower than on the apple skin, which indicates a better wettability of persimmon with these types of formulations. The values depended on the coating composition, with the highest contact angle corresponding to the S:G formulation in apples. This could imply problems for the extension of this coating on the apple surface. A previous study [29] reported that 10^5 mg·L^{-1} of Tween 85 must be added to ensure the S:G coating spreadability on the apple surface, whereas no surfactant was necessary to enhance the CFS spreadability when these contained emulsified or lecithin-encapsulated EO. Therefore, Tween 85 was added to the S:G formulation and tested in a preliminary trial with apples, in comparison with the CFS without Tween 85, in order to analyse the effect of the surfactant on the fruit quality during storage, as discussed in the next section.

Table 1. Rheological parameters (flow behavior index, n; consistency index, K; apparent viscosity at $100\ s^{-1}$, η) and contact angle (θ) of the coating forming solution (CFS) on the skin of 'Golden Delicious' apple and 'Rojo Brillante' persimmon. Mean values and standard deviations.

CFS	Rheological Behavior			Contact Angle (θ)	
	n	K (mPa·s) ^{n}	η at $100\ s^{-1}$ (mPa·s)	Apple	Persimmon
S:G	0.854 ± 0.001 [d]	65.0 ± 0.2 [a]	33.1 ± 0.1 [c]	96 ± 2 [e]	67 ± 3 [cd]
S:G-L	0.74 ± 0.01 [a]	114 ± 3 [b]	35.0 ± 0.1 [d]	85 ± 3 [d]	72 ± 2 [e]
S:G-0.25	0.86 ± 0.01 [d]	59 ± 9 [a]	31 ± 3 [b]	69 ± 3 [a]	65 ± 3 [c]
S:G-0.25-L	0.815 ± 0.001 [c]	59.7 ± 0.5 [a]	25.5 ± 0.3 [a]	73 ± 2 [b]	50 ± 6 [a]
S:G-0.5	0.766 ± 0.004 [b]	124 ± 3 [c]	42.2 ± 0.2 [e]	77 ± 2 [c]	68 ± 2 [d]
S:G-0.5-L	0.809 ± 0.002 [c]	60 ± 1 [a]	25.05 ± 0.03 [a]	74 ± 2 [b]	55 ± 4 [b]

Different superscript letters within the same column indicate significant differences among CFS according to Fisher's least significant difference (LSD) test ($p < 0.05$).

3.2. Effect of the Incorporation of Tween 85 into CFS on Apple Quality

The incorporation of Tween 85 into the S:G CFS significantly decreased the contact angle on the apple surface (from 96 ± 2 to 47 ± 3) and increased the apparent viscosity (from 37.9 ± 0.3 to 187.8 ± 0.1 mPa·s). As expected, both changes affected the retention/adhesion of the CFS on the apple surface, as shown in the values of SDS on the fruit, which ranged from 2.6 ± 0.8 to 3.4 ± 0.5 g·m^{-2}. Interactions of Tween 85 with the CFS components and with the fruit surface affected both the viscosity and the contact angle of the CFS. As described by Marín et al. [32], surfactant molecules form complexes, with the helical conformation of amylose favouring the chain aggregation and increasing the system viscosity. Likewise, this complex formation implies that a high amount of surfactant is required to enhance the spreading of the CFS on the fruit surface, as discussed by Sapper et al. [29]. The increase in the SDS values for the S:G formulation with Tween 85 can be attributed to the higher solid content of the formulation, the greater viscosity that limits liquid gravitational drainage, and the lower contact angle. However, given the amphiphilic nature of the surfactant, its interactions with the natural wax of the fruit cuticle could also modify the overall barrier properties of the wax-coating assembly on the fruit surface. As is known, cuticular waxes are the primary components of the cuticle that are responsible for its permeability and wettability. These waxes are embedded in the cutin and form a continuous layer on the top of the cutin [35]. It has been reported that the cuticular wax content in apple fruit increases during fruit development and storage [36].

Table 2 shows the relative weight loss rate, respiration rate, and puncture parameters of apples after 7 days of storage at 25 °C for samples coated with the S:G formulation containing Tween 85 or not, in comparison with the uncoated control sample. Little differences in the relative weight loss rate were observed between the uncoated control sample and the one coated with the surfactant-free formulation. However, a significantly higher weight loss rate was observed for those coated with the formulation containing Tween 85. The coatings reduced the O_2 consumption rate of the fruit, which can be attributed to the low O_2 permeability of these films [24], but this reduction was particularly significant for the coating containing Tween 85. The CO_2 production rate was not significantly affected by the S:G coating compared to the control sample, but the coating containing Tween 85 significantly reduced this rate. As a consequence, the respiratory quotient was higher than 1 for both coated samples, indicating the creation of a modified atmosphere in the fruit and a shift towards anaerobic respiration pathways. The incorporation of Tween 85 resulted in a general decrease in the gas transfer rate and an increase in water transfer rate. As discussed above, the interactions of Tween 85 with the cuticular waxes, as well as its effect on the decrease in the cohesion forces of the S:G matrix (limiting of chain packing), could explain the changes observed in the gas and water vapour barrier properties of the coating and their effect on the fruit. These changes also had an effect on the fruit texture, as shown in Table 2. Although all the samples exhibited similar fruit firmness as deduced from the lack of significant differences among treatments regarding the maximum puncture force, there were significant differences in the maximum penetration distance (d_{max}) at the tissue rupture. Fruits coated with the S:G formulation containing Tween 85 had significantly higher d_{max} values, which reflect changes in the tissue texture. This fact can be related to the greater loss of water, and therefore, cellular turgidity, which is associated with a more marked superficial dehydration of the fruit with this coating. This factor is considered one of the main causes of texture changes in fruit [37]. After 7 days of storage, all the samples had higher d_{max} than the fruit at the initial time, with those coated with the CFS containing Tween 85 being significantly more deformable. Therefore, the use of Tween in the S:G formulation to improve its wettability on the apple surface was discarded on the basis of the negative effects on the fruit weight loss and texture.

Table 2. Effect of the incorporation of Tween 85 into the cassava starch:low acyl gellan gum (S:G) coating formulation on the postharvest behavior and quality of coated 'Golden Delicious' apples: relative weight loss rate (day^{-1}), respiration rates (consumption of O_2 and production of CO_2, mL·kg^{-1}·h^{-1}), respiration quotient (RQ), and values of the maximum puncture force (F_{max}, N) and penetration distance (d_{max}, mm) after 7 days of storage at 25 °C. Uncoated samples were used for values at harvest and the control after 7 days of storage.

	Control	Control	S:G	S:G-Tween 85
	Initial Time	**7 days**		
Weight loss rate	–	0.36 ± 0.02 [a]	0.36 ± 0.01 [a]	0.66 ± 0.06 [b]
F_{max}	43 ± 7	46 ± 6 [a]	49 ± 8 [a]	46 ± 4 [a]
d_{max}	3.0 ± 0.3	3.6 ± 0.4 [a]	3.9 ± 0.5 [a]	4.6 ± 0.7 [b]
R O_2	12.94 ± 0.05	12.9 ± 1.3 [b]	11.4 ± 0.8 [b]	7.77 ± 0.02 [a]
R CO_2	13.9 ± 0.6	15.2 ± 0.7 [b]	18 ± 1 [c]	11.0 ± 0.2 [a]
RQ	1.07 ± 0.05	1.18 ± 0.07 [a]	1.58 ± 0.03 [c]	1.41 ± 0.03 [b]

Different superscript letters within the same row indicate significant differences among CFS according to Fisher's LSD test ($p < 0.05$).

3.3. Effect of CFS on Postharvest Behaviour and Quality of Apples and Persimmons

Table 3 shows the initial values of respiration rates and puncture parameters of apples and persimmons, and Table 4 shows the same parameters, together with the values of SDS, for coated and uncoated 'Golden Delicious' apples and 'Rojo Brillante' persimmons after storage at 25 °C. SDS values are indicators of the coating thickness on the fruit; the higher the SDS, the thicker the coating. The SDS value depends on the amount of CFS that adhered to the surface of the fruit and the total solid content of the formulation. The former, in turn, is affected by the wetting/spreading capacity and the viscosity of the coating formulations. In general, the SDS values were higher in apples than in persimmons, which could be related to differences in both the surface tension of the skin [29] and in the skin morphology of the fruit. Thus, persimmons are characterized by a smooth skin, where the lack of small superficial pores could limit the capillary retention of the liquid fraction. Similarly, the CFS composition slightly affected the SDS differently depending on the fruit. In apples, the presence of lecithin in the formulation significantly reduced the SDS, whereas smaller differences associated with the CFS composition were observed in persimmons, and these were seemingly more closely related to the solid content of the CFS (incorporation of EO and/or lecithin to the formulations).

Table 3. Respiration rates (consumption of O_2 and production of CO_2, mL·kg^{-1}·h^{-1}), respiration quotient (RQ), and maximum puncture force (F_{max}, N) and penetration distance (d_{max}, mm) of uncoated 'Golden Delicious' apples and 'Rojo Brillante' persimmons at initial time. Mean values and standard deviations.

	R O_2	R CO_2	RQ	F_{max}	d_{max}
Apple	13.7 ± 1.5	12.2 ± 0.2	0.9 ± 0.1	29 ± 2	2.2 ± 0.1
Persimmon	5.7 ± 1.1	5.4 ± 0.9	0.94 ± 0.02	21.1 ± 0.5	2.5 ± 0.3

Table 4. Surface density of solids (SDS, g·m^{-2}), weight loss rate (day^{-1}), respiration rates (consumption of O_2 and production of CO_2, mL·kg^{-1}·h^{-1}), respiration quotient (RQ), maximum puncture force (F_{max}, N), and penetration distance (d_{max}, mm) of coated and uncoated 'Golden Delicious' apples and 'Rojo Brillante' persimmons for 7 or 14 days of storage at 25 °C.

	Control	S:G	S:G-L	S:G-0.25	S:G-0.25-L	S:G-0.5	S:G-0.5-L
				Apple			
SSD	–	1.3 ± 0.3 [c]	0.8 ± 0.1 [a]	1.4 ± 0.3 [c]	1.1 ± 0.1 [b]	1.5 ± 0.2 [c]	1.0 ± 0.1 [b]
Weight loss rate (14 d)	0.23 ± 0.03 [a]	0.21 ± 0.03 [a]	0.20 ± 0.05 [a]	0.21 ± 0.03 [a]	0.20 ± 0.03 [a]	0.20 ± 0.03 [a]	0.22 ± 0.03 [a]
R O_2 (7 d)	6.0 ± 1.4 [ab]	6.0 ± 1.5 [ab]	6.5 ± 1.6 [ab]	6.6 ± 0.8 [ab]	6.5 ± 1.0 [ab]	7.6 ± 0.6 [b]	5.1 ± 0.9 [a]
R CO_2 (7 d)	6.5 ± 1.2 [a]	7.8 ± 0.8 [ab]	8.4 ± 2.0 [ab]	8.8 ± 0.1 [ab]	8.3 ± 0.7 [ab]	9.7 ± 1.0 [b]	6.8 ± 0.9 [a]
RQ (7 d)	1.1 ± 0.1 [a]	1.3 ± 0.1 [b]	1.29 ± 0.04 [b]	1.4 ± 0.1 [b]	1.3 ± 0.1 [b]	1.3 ± 0.1 [b]	1.3 ± 0.1 [b]
R O_2 (14 d)	5.3 ± 0.8 [a]	8.4 ± 1.1 [d]	7.3 ± 0.5 [cd]	7.0 ± 0.1 [bcd]	5.9 ± 0.8 [ab]	6.5 ± 0.9 [abc]	5.6 ± 0.9 [ab]
R CO_2 (14 d)	6.2 ± 1.3 [a]	10.4 ± 1.1 [d]	9.0 ± 0.9 [cd]	8.1 ± 0.2 [bc]	7.2 ± 0.6 [ab]	7.9 ± 0.6 [bc]	7.2 ± 0.6 [ab]
RQ (14 d)	1.2 ± 0.1 [a]	1.2 ± 0.1 [a]	1.23 ± 0.05 [a]	1.2 ± 0.1 [a]	1.2 ± 0.1 [a]	1.2 ± 0.1 [a]	1.3 ± 0.1 [a]
F_{max} (14 d)	27 ± 2 [ab]	31 ± 3 [c]	30 ± 4 [bc]	29 ± 3 [abc]	30.4 ± 1.5 [bc]	32 ± 2 [c]	26.1 ± 1.2 [a]
d_{max} (14 d)	3 ± 1 [a]	3 ± 1 [a]	3 ± 1 [a]	3 ± 0 [a]	3 ± 1 [a]	3 ± 1 [a]	3.4 ± 0.5 [a]
				Persimmon			
SSD	–	0.5 ± 0.1 [a]	0.7 ± 0.1 [ab]	0.7 ± 0.2 [b]	0.8 ± 0.1 [bc]	0.9 ± 0.1 [c]	0.7 ± 0.2 [b]
Weight loss rate (14 d)	0.7 ± 0.1 [b]	0.6 ± 0.1 [ab]	0.6 ± 0.1 [ab]	0.52 ± 0.03 [a]	0.56 ± 0.06 [ab]	0.6 ± 0.1 [ab]	0.7 ± 0.1 [b]
R O_2 (7 d)	5.2 ± 0.3 [a]	3.3 ± 1.6 [a]	3.8 ± 0.6 [a]	2.9 ± 0.2 [a]	3.9 ± 2.1 [a]	3.6 ± 0.3 [a]	4.1 ± 1.7 [a]
R CO_2 (7 d)	5.6 ± 0.5 [a]	3.8 ± 1.0 [a]	4.2 ± 0.9 [a]	3.4 ± 0.4 [a]	4.5 ± 2.0 [a]	4.6 ± 0.7 [a]	4.8 ± 2.1 [a]
RQ (7 d)	1.08 ± 0.04 [a]	1.2 ± 0.3 [a]	1.11 ± 0.05 [a]	1.19 ± 0.03 [a]	1.3 ± 0.3 [a]	1.3 ± 0.1 [a]	1.2 ± 0.1 [a]
R O_2 (14 d)	3.7 ± 0.6 [a]	4.2 ± 1.8 [a]	3.0 ± 1.1 [a]	3.3 ± 0.4 [a]	2.2 ± 0.2 [a]	2.6 ± 0.6 [a]	3.5 ± 1.9 [a]
R CO_2 (14 d)	3.9 ± 1.1 [a]	5.3 ± 2.6 [a]	4.1 ± 1.0 [a]	3.9 ± 1.8 [a]	2.2 ± 0.2 [a]	2.9 ± 0.4 [a]	4.2 ± 1.3 [a]
RQ (14 d)	1.0 ± 0.1 [a]	1.3 ± 0.1 [a]	1.4 ± 0.2 [a]	1.2 ± 0.4 [a]	1.0 ± 0.0 [a]	1.1 ± 0.1 [a]	1.3 ± 0.3 [a]
F_{max} (14 d)	21 ± 4 [a]	27 ± 3 [bc]	23 ± 4 [ab]	27 ± 5 [bc]	28 ± 6 [c]	29 ± 6 [c]	28 ± 6 [c]
d_{max} (14 d)	5 ± 1 [ab]	6 ± 1 [bc]	5 ± 1 [a]	5 ± 1 [a]	5.2 ± 0.4 [abc]	6 ± 1 [c]	6 ± 1 [bc]

Different superscript letters within the same row indicate significant differences among CFS according to Fisher's LSD test ($p < 0.05$).

The rate of relative weight loss after 14 days of storage was not significantly affected by coating application or composition, and ranged between 0.20–0.23 day^{-1}. In persimmons, water loss rates were higher than in apples and varied depending on the coating formulation. The highest values (0.7 day^{-1}) were obtained for uncoated samples, and those coated with CFS containing the highest content of lecithin-encapsulated EO (maximum lipid content in the film) and the lowest value (0.52 day^{-1}) was obtained for samples coated with the formulation with emulsified EO (without lecithin) at the lowest ratio (minimum lipid content in the film). No significant differences were found between the other coating formulations and the control samples. These results indicate that persimmon fruits were more sensitive to dehydration than apples under these storage conditions, and the coatings with the lowest ratio of emulsified EO exerted a protective effect. As mentioned above, apples and persimmons are naturally covered by a continuous wax layer that provides the resistance to water movement across the cuticle. The differences in the water resistance of the untreated fruits can be attributed to the particular fruit physiology, skin morphology, and the composition of the natural waxes. The application of coatings containing hydrophobic compounds should improve the moisture resistance of the fruit, as an additional layer is deposited over the natural waxes. In the present study, none of the coatings reduced weight loss in apples, and only the coating that had the lowest amount of EO and no lecithin prevented water loss in persimmons. This might indicate a partial removal and/or modification of the natural waxes that are present on the peel of the fruits, resulting in no reduction in weight loss; so, further studies should be conducted in order to understand the effect of the EO and lecithin on the water barrier properties of coated apples and persimmons. Some other studies also reflected no effect of coatings based on biopolymers and lipids on the weight loss reduction of different fruits, such as apples [31], plums [38], table grapes [39], or cherry tomatoes [40], compared to uncoated fruits.

Table 4 also shows respiration rates of both fruit at 7 and 14 days of storage. In apples, a multifactor ANOVA (results not shown) did not reveal a significant effect of the storage time on respiration rates, although the coatings had a significant influence. In general, coatings tend to increase the O_2 consumption and CO_2 production rates with respect to the control sample, and only those containing lecithin-encapsulated EO showed no significant differences with respect to the control sample. The alterations in the respiration pathway affected the RQ, which indicates the nature of the substrate used during the respiration process. Thus, a RQ equal to 1.0 indicates that the metabolic substrates are carbohydrates, whereas an RQ higher than 1 indicates that the substrates are organic acids [41]. The multifactor ANOVA in RQ reveals a significant effect of storage time and coating formulation. RQ slightly increased at 14 days, and was higher in all the coated samples. This indicates that the metabolic substrates are shifting from carbohydrates to organic acids [41] more quickly in coated samples.

In contrast, coatings were observed to have no significant effect on the respiration rates of persimmons, which exhibited lower respiration rates than apples, with a respiration quotient of nearly 1. Climacteric fruit, such as apples, exhibit a peak of respiration and ethylene (C_2H_4) production associated with senescence or ripening [12], which could explain the observed differences.

The texture changes in fruits depend on both cell wall degradation and the loss of tissue turgidity [42]. Table 4 shows the values of the F_{max} and d_{max} for the different coated and uncoated fruits after 14 days of storage. No significant changes in the maximum puncture force (failure point) were observed for either uncoated or coated apples after storage with respect to the initial values before storage (Tables 3 and 4). However, although no significant differences were observed in terms of the maximum penetration distance between coated and control apples at the end of the storage, the values were slightly higher than before storage, which can be associated with a loss of cellular turgidity due to the superficial dehydration of the apples. The limited water vapor barrier capacity of these films [24] and their relative lack of thickness on the fruit mean that these are scarcely effective at controlling moisture transfer in apples.

In the case of persimmons, the coatings had a significant effect ($p < 0.05$), maintaining the firmness of the fruit. On the other hand, although there were no notable differences in terms of the maximum penetration distance at the failure point (5–6 mm) between coated and uncoated samples at the end of the storage, the values were significantly higher than the initial value (2.5 mm) before storage. This indicates changes in the texture of the tissue over time, which can be related to the progress in maturity and water loss. The F_{max} values increased from 21.1 up to 31 N, which could be attributed to the greater deformability of the tissue allowing for deeper penetration without failure, thus accumulating more compressive and shear resistance [43]. The smallest changes occurred in the sample coated with CFS containing lecithin without EO.

The above results show that coatings have a different effect depending on whether the fruit is an apple or persimmon, which can be attributed to the different physiological patterns of the fruits and the specific interactions with the coatings. Although respiration patterns were slightly modified by coatings on apples with no effect on water loss, coatings exerted a better control of water loss in persimmon; however, these did not maintain the firmness mainly due to the progress of fruit ripening.

3.4. Fungal Decay

Table 5 shows the development of fungal decay on artificially inoculated 'Golden Delicious' apples and 'Rojo Brillante' persimmons. The applications of starch–gellan coatings did not significantly reduce the disease incidence on apples inoculated with *B. cinerea*, as compared to non-coated ones (control) after 7 or 12 days of storage at 20 °C. No effect of the addition of EO was observed, despite what had been observed in a prior in vitro study, where starch–gellan films with thyme EO exhibited a marked antifungal effect [24]. Nevertheless, all the coatings, regardless of their composition, significantly reduced the severity of gray mold with respect to the control samples (20–30% reduction), with no particular observed effect of the antifungal EO.

Table 5. Mean values and standard deviations of disease incidence and severity of gray mold on 'Golden Delicious' apples artificially inoculated with *Botrytis cinerea* and black spot on 'Rojo Brillante' persimmons artificially inoculated with *Alternaria alternata*. Fruit were coated 24 h after fungal inoculation and incubated at 20 °C and 85% RH for 7 and 12 days. Mean values of the reduction in disease incidence and severity are also shown.

	Disease Incidence (%)		Reduction of Incidence (%)		Disease Severity (mm)		Reduction of Severity (%)	
	7 Days	12 Days	7 Days	12 Days	7 Days	12 Days	7 Days	12 Days
Apple gray mold								
Control	100 ± 0 [a]	100 ± 0 [a]	–	–	70 ± 5 [b]	100 ± 5 [b]	–	–
S:G	75 ± 25 [a]	83 ± 14 [a]	25	17	44 ± 5 [a]	74 ± 13 [a]	32	26
S:G-L	92 ± 14 [a]	92 ± 14 [a]	8	8	47 ± 11 [a]	73 ± 21 [a]	27	27
S:G-0.25	75 ± 25 [a]	83 ± 14 [a]	25	17	53 ± 8 [ab]	76 ± 15 [a]	19	24
S:G-0.25-L	75 ± 25 [a]	75 ± 25 [a]	25	25	44 ± 11 [a]	64 ± 17 [a]	33	36
S:G-0.5	92 ± 14 [a]	92 ± 14 [a]	8	8	45 ± 10 [a]	81 ± 7 [ab]	32	19
S:G-0.5-L	100 ± 0 [a]	100 ± 0 [a]	0	0	47 ± 1 [a]	69 ± 10 [a]	29	31
Persimmon black spot								
Control	68 ± 3 [b]	73 ± 5 [bc]	–	–	10.6 ± 0.8 [a]	21.9 ± 1.9 [b]	–	–
S:G	38 ± 9 [a]	45 ± 12 [a]	44	39	9.3 ± 1.5 [a]	15.0 ± 3.0 [a]	12	32.9
S:G-L	70 ± 10 [b]	78 ± 7 [bc]	0	0	9.7 ± 1.2 [a]	17.0 ± 3.0 [a]	9	20.8
S:G-0.25	42 ± 7 [a]	57 ± 10 [ab]	39	23	10.8 ± 1.1 [a]	15.5 ± 1.4 [a]	0	29.4
S:G-0.25-L	58 ± 8 [ab]	72 ± 6 [bc]	14	2	12.7 ± 0.6 [a]	18.8 ± 1.9 [ab]	0	14.2
S:G-0.5	42 ± 7 [a]	53 ± 7 [ab]	39	27	11.4 ± 0.3 [a]	20.2 ± 0.5 [ab]	0	8
S:G-0.5-L	72 ± 7 [b]	82 ± 8 [c]	0	0	10.4 ± 0.8 [a]	17.6 ± 0.6 [a]	2	20

For each disease, different superscript letters within the same column indicate significant differences among CFS according to Fisher's LSD test ($p < 0.05$).

Starch–gellan coatings were more effective at reducing the incidence of black spot caused by *A. alternata* on persimmons (up to 40% reduction), although coatings containing lecithin were not effective, and the presence of EO was not observed to have any significant effect. Disease severity was not significantly reduced in coated persimmons. A multifactorial analysis (factors: the presence of lecithin and EO concentration) revealed two things: there was no significant influence of the EO, and the lecithin had a negative effect on the reduction of disease incidence and severity in infected fruit.

In general, applying a coating had a positive antifungal effect both on apples (a significant reduction in the severity of gray mold) and persimmons (a significant reduction in the incidence of black spot), but this antifungal effect was milder than that observed in in vitro work with EO incorporated into the same type of films. Similar behavior has recently been reported by da Rocha Neto et al. [15] for apples. They observed a complete inhibition of the in vitro germination of *P. expansum* by using melaleuca, palmarosa, and star anise EOs in vapor phase, but these treatments had only a minor effect on inoculated apples, regardless of the EO used. As previously reported [22,44], this indicates that the in vivo effectiveness of EOs cannot be anticipated by their antifungal activity in in vitro tests, and that interactions between EOs and fungal pathogens are modulated by the fruit host and the conditions in the infection court, often resulting in reduced disease control ability. An important difference in the potential effect of EO with respect to in vitro tests could be related with the degree of coating plasticization, which may affect the release of EO. In in vitro tests, films are directly applied on the wet culture medium, whereas coatings are applied on the dried fruit surface. This fact could limit the release of the active compounds from the polymer matrix, hindering their antifungal action.

Likewise, EO compounds may also affect some physiological changes in the fruit, which could decrease the fruit's natural defenses against the fungal attack. The generally negative effect of lecithin could also be attributed to the lipid interactions with the fruit's waxy coatings, which could also weaken the natural resistance to disease, counteracting the induced coating protection. Other surfactant lipids, such as Tween 85, also seemed to exert a negative effect on the barrier capacity of the natural wax-coating assembly, as observed in apples. The gas exchange on the fruit surface could also play an important role in postharvest disease development, which could explain the generally positive effect of coatings at reducing fungal growth on infected fruit. Therefore, interactions of coatings and their components with the fruit surface always constitute a distinguishing factor to define the particular behavior of coated fruit [45], and in the case of coatings formulated with antifungal ingredients, these interactions can affect the in vivo disease control ability of the coating [46]. The results obtained in the present study with the addition of thyme EO to starch-based coatings were not anticipated. Numerous previous studies have shown that the formulation of antifungal films and coatings with EOs either provided disease control ability or increased that of the coating alone due to an important synergistic effect against various important postharvest pathogens, including *B. cinerea* and *A. alternata* [12,47,48]. However, this is not always the case, and other reports showed no significant benefit gained from the addition of EOs [49]. It seems that a wide variability in disease control efficacy can be observed, which is basically due to the numerous factors that can influence the antifungal properties of films and coatings. The following can be cited among the most important: nature of the composite matrix of the coating; type and concentration of the antifungal compound(s); species and strain of the target postharvest pathogen; species, cultivar, and physical and physiological condition of the fruit host; and postharvest environmental conditions [46]. Therefore, further studies would be required to analyze the influence of these factors.

4. Conclusions

Starch–gellan coatings containing or not emulsified or lecithin-encapsulated EO had a different effect on the postharvest parameters (weight loss, respiration rates and firmness changes) when applied on apples and persimmons, depending on the coating composition and type of fruit. None of the coating formulations reduced the weight loss in apples, although these prevented water loss in persimmons. In contrast, although the coating was not observed to have any significant effect on the respiration rates and respiration quotient of persimmons, the respiration rates and quotient in apples were promoted. Coatings did not affect the changes in fruit firmness in apples or persimmons; nevertheless, in the latter, these may be mainly associated with the ripening progress. Regarding fungal decay, coatings without lecithin reduced the incidence of black spot caused by *A. alternata* in persimmons, regardless of the thyme EO content. Likewise, these reduced the severity of gray mold caused by *B. cinerea* infection in apple. The addition of EO did not exert an antifungal effect in the fruit despite its proven antifungal action in previous in vitro tests. Therefore, the particular characteristics of the fruit and the interactions in the infection site (peel wounds) seriously affected the in vivo effectiveness of coatings of certain composition. No positive effect of lecithin was observed on the controlled postharvest parameters affecting fruit quality and physiological behavior in either apples or persimmons; EO did not exert additional antifungal action and seemed to exert a negative effect on some other fruit quality attributes. Then, starch–gellan coatings without lecithin or thyme EO demonstrated the potential to be used in persimmons in order to control weight loss and reduce the incidence of infections caused by *A. alternata*.

Author Contributions: Conceptualization, M.S., L.P., M.B.P.-G., and A.C.; methodology, L.P., M.B.P.-G., and A.C.; investigation, M.S.; writing—original draft preparation, M.S. and A.C.; writing—review and editing, M.S., L.P., M.B.P.-G. and A.C.

Funding: This research was funded by the "Ministerio de Economía y Competitividad" (MINECO) of Spain through the projects AGL2016-76699-R and RTA2015-00037-C02. Mayra Sapper thanks the "Conselleria de Educación, Investigación, Cultura y Deporte de la Generalitat Valenciana" for the Santiago Grisolía grant GRISOLIA/2015/001.

Conflicts of Interest: The authors declare no conflict of interest.

References

1. Singh, D.; Sharma, R.R. Postharvest diseases of fruits and vegetables and their management. In *Postharvest Disinfection of Fruits and Vegetables*; Siddiqui, M.W., Ed.; Academic Press: London, UK, 2018; pp. 1–52. [CrossRef]

2. Sivakumar, D.; Bautista-Baños, S. A review on the use of essential oils for postharvest decay control and maintenance of fruit quality during storage. *Crop Prot.* **2014**, *64*, 27–37. [CrossRef]

3. Troncoso-Rojas, R.; Tiznado-Hernández, M.E. Alternaria alternata (black rot, black spot). In *Postharvest Decay: Control Strategies*; Bautista-Baños, S., Ed.; Academic Press: London, UK, 2014; pp. 147–187. [CrossRef]

4. Batta, Y.A. Postharvest biological control of apple gray mold by *Trichoderma harzianum* Rifai formulated in an invert emulsion. *Crop Prot.* **2004**, *23*, 19–26. [CrossRef]

5. Palou, L.; Montesinos-Herrero, C.; Tarazona, I.; Besada, C.; Taberner, V. Incidence and etiology of postharvest fungal diseases of persimmon (*Diospyros Kaki* Thunb. Cv. Rojo Brillante) in Spain. *Plant Dis.* **2015**, *99*, 1416–1425. [CrossRef] [PubMed]

6. Ma, L.; He, J.; Liu, H.; Zhou, H. The phenylpropanoid pathway affects apple fruit resistance to *Botrytis cinerea*. *J. Phytopathol.* **2018**, *166*, 206–215. [CrossRef]

7. Romanazzi, G.; Feliziani, E. Botrytis cinerea (gray mold). In *Postharvest Decay: Control Strategies*; Bautista-Baños, S., Ed.; Academic Press: London, UK, 2014; pp. 131–146. [CrossRef]

8. Prusky, D.; Eshel, D.; Kobiler, I.; Yakoby, N.; Beno-Moualem, D.; Ackerman, M.; Zuthji, Y.; Ben Arie, R. Postharvest chlorine treatments for the control of the persimmon black spot disease caused by *Alternaria alternata*. *Postharvest Biol. Technol.* **2001**, *22*, 271–277. [CrossRef]

9. Biton, E.; Kobiler, I.; Feygenberg, O.; Yaari, M.; Kaplunov, T.; Ackerman, M.; Friedman, H.; Prusky, D. The mechanism of differential susceptibility to alternaria black spot, caused by *Alternaria alternata*, of stem-and bottom-end tissues of persimmon fruit. *Postharvest Biol. Technol.* **2014**, *94*, 74–81. [CrossRef]

10. Olivas, G.; Barbosa-Cánovas, G. Edible films and coatings for fruits and vegetables. In *Edible Films and Coatings for Food Applications*; Embuscado, M.E., Huber, K.C., Eds.; Springer: New York, NY, USA, 2009; pp. 211–244. [CrossRef]

11. Maftoonazad, N.; Ramaswamy, H.S.; Marcotte, M. Shelf-life extension of peaches through sodium alginate and methyl cellulose edible coatings. *Int. J. Food Sci. Technol.* **2008**, *43*, 951–957. [CrossRef]

12. Fonseca, S.C.; Oliveira, F.A.R.; Brecht, J.K. Modelling respiration rate of fresh fruits and vegetables for modified atmosphere packages: A review. *J. Food Eng.* **2002**, *52*, 99–119. [CrossRef]

13. Morice, I.M.; Shorland, F.B. Composition of the surface waxes of apple fruits and changes during storage. *J. Sci. Food Agric.* **1973**, *24*, 1331–1339. [CrossRef]

14. Prusky, D.; Kobiler, I.; Akerman, M.; Miyara, I. Effect of acidic solutions and acidic prochloraz on the control of postharvest decay caused by *Alternaria alternata* in mango and persimmon fruit. *Postharvest Biol. Technol.* **2006**, *42*, 134–141. [CrossRef]

15. Da Rocha Neto, A.C.; Navarro, B.B.; Canton, L.; Maraschin, M.; Di Piero, R.M. Antifungal activity of palmarosa (*Cymbopogon martinii*), tea tree (*Melaleuca alternifolia*) and star anise (*Illicium verum*) essential oils against *Penicillium expansum* and their mechanisms of action. *LWT-Food Sci. Technol.* **2019**, *105*, 385–392. [CrossRef]

16. El Ouadi, Y.; Manssouri, M.; Bouyanzer, A.; Majidi, L.; Bendaif, H.; Elmsellem, H.; Shariati, M.A.; Melhaoui, A.; Hammouti, B. Essential oil composition and antifungal activity of *Melissa officinalis* originating from North-Est Morocco, against postharvest phytopathogenic fungi in apples. *Microb. Pathog.* **2017**, *107*, 321–326. [CrossRef] [PubMed]

17. Desjobert, J.M.; Cristofari, G.; Paolini, J.; Costa, J.; Majidi, L.; Znini, M. Essential oil composition and antifungal activity of *Pulicaria mauritanica* coss., against postharvest phytopathogenic fungi in apples. *LWT-Food Sci. Technol.* **2013**, *54*, 564–569. [CrossRef]

18. Perdones, A.; Sánchez-González, L.; Chiralt, A.; Vargas, M. Effect of chitosan-lemon essential oil coatings on storage-keeping quality of strawberry. *Postharvest Biol. Technol.* **2012**, *70*, 32–41. [CrossRef]

19. Daniel, C.K.; Lennox, C.L.; Vries, F.A. In vivo application of garlic extracts in combination with clove oil to prevent postharvest decay caused by *Botrytis cinerea*, *Penicillium expansum* and *Neofabraea alba* on apples. *Postharvest Biol. Technol.* **2014**, *99*, 88–92. [CrossRef]

20. Munhuweyi, K.; Caleb, O.J.; Lennox, C.L.; van Reenen, A.J.; Opara, U.L. In vitro and in vivo antifungal activity of chitosan-essential oils against pomegranate fruit pathogens. *Postharvest Biol. Technol.* **2017**, *129*, 9–22. [CrossRef]

21. Bakkali, F.; Averbeck, S.; Averbeck, D.; Idaomar, M. Biological effects of essential oils-A review. *Food Chem. Toxicol.* **2008**, *46*, 446–475. [CrossRef]

22. Palou, L.; Ali, A.; Fallik, E.; Romanazzi, G. GRAS, plant-and animal-derived compounds as alternatives to conventional fungicides for the control of postharvest diseases of fresh horticultural produce. *Postharvest Biol. Technol.* **2016**, *122*, 41–52. [CrossRef]

23. Sánchez-González, L.; Vargas, M.; González-Martínez, C.; Chiralt, A.; Cháfer, M. Use of essential oils in bioactive edible coatings: A review. *Food Eng. Rev.* **2011**, *3*, 1–16. [CrossRef]

24. Sapper, M.; Wilcaso, P.; Santamarina, M.P.; Roselló, J.; Chiralt, A. Antifungal and functional properties of starch-gellan films containing thyme (*Thymus zygis*) essential oil. *Food Control.* **2018**, *92*, 505–515. [CrossRef]

25. Plotto, A.; Roberts, D.D.; Roberts, R.G. Evaluation of plant essential oils as natural postharvest disease control of tomato (*Lycopersicon esculentum*). *Acta Hortic.* **2003**, *628*, 737–745. [CrossRef]

26. Cano, A.; Fortunati, E.; Cháfer, M.; Kenny, J.M.; Chiralt, A.; González-Martínez, C. Properties and ageing behaviour of pea starch films as affected by blend with poly(vinyl alcohol). *Food Hydrocoll.* **2015**, *48*, 84–93. [CrossRef]

27. Jiménez, A.; Fabra, M.J.; Talens, P.; Chiralt, A. Edible and biodegradable starch films: A review. *Food Bioprocess Technol.* **2012**, *5*, 2058–2076. [CrossRef]

28. Valencia-Sullca, C.; Jiménez, M.; Jiménez, A.; Atarés, L.; Vargas, M.; Chiralt, A. Influence of liposome encapsulated essential oils on properties of chitosan films. *Polym. Int.* **2016**, *65*, 979–987. [CrossRef]

29. Sapper, M.; Bonet, M.; Chiralt, A. Wettability of starch-gellan coatings on fruits, as affected by the incorporation of essential oil and/or surfactants. *LWT-Food Sci. Technol.* under review.

30. Kwok, D.Y.; Neumann, A.W. Contact angle measurement and contact angle interpretation. *Adv. Colloid Interface Sci.* **1999**, *81*, 167–249. [CrossRef]

31. Bai, J.; Baldwin, E.A.; Hagenmaier, R.H. Alternatives to shellac coatings provide comparable gloss, internal gas modification, and quality for "delicious" apple fruit. *HortScience* **2002**, *37*, 559–563. [CrossRef]

32. Marín, A.; Atarés, L.; Cháfer, M.; Chiralt, A. Properties of biopolymer dispersions and films used as carriers of the biocontrol agent *Candida sake* CPA-1. *LWT-Food Sci. Technol.* **2017**, *79*, 60–69. [CrossRef]

33. Castelló, M.L.; Fito, P.J.; Chiralt, A. Changes in respiration rate and physical properties of strawberries due to osmotic dehydration and storage. *J. Food Eng.* **2010**, *97*, 64–71. [CrossRef]

34. Saei, A.; Tustin, D.S.; Zamani, Z.; Talaie, A.; Hall, A.J. Cropping effects on the loss of apple fruit firmness during storage: the relationship between texture retention and fruit dry matter concentration. *Sci. Hortic.* **2011**, *130*, 256–265. [CrossRef]

35. Belding, R.; Blankenship, S.; Young, E.; Leidy, R. Composition and variability of epicuticular waxes in apple cultivars. *J. Am. Soc. Hortic. Sci.* **1998**, *123*, 348–356. [CrossRef]

36. Ju, Z.; Bramlage, W.J. Developmental changes of cuticular constituents and their association with ethylene during fruit ripening in "Delicious" apples. *Postharvest Biol. Technol.* **2001**, *21*, 257–263. [CrossRef]

37. Saberi, B.; Golding, J.B.; Marques, J.R.; Pristijono, P.; Chockchaisawasdee, S.; Scarlett, C.J.; Stathopoulos, C.E. Application of biocomposite edible coatings based on pea starch and guar gum on quality, storability and shelf life of 'Valencia' oranges. *Postharvest Biol. Technol.* **2018**, *137*, 9–20. [CrossRef]

38. Navarro-Tarazaga, M.L.; Sothornvit, R.; Pérez-Gago, M.B. Effect of plasticizer type and amount on hydroxypropyl methylcellulose-beeswax edible film properties and postharvest quality of coated plums (cv. Angeleno). *J. Agric. Food Chem.* **2008**, *56*, 9502–9509. [CrossRef] [PubMed]

39. Pastor, C.; Sanchez-Gonzalez, L.; Marcilla, A.; Chiralt, A.; Cháfer, M.; Gonzalez-Martinez, C. Quality and safety of table grapes coated with hydroxypropylmethylcellulose edible coatings containing propolis extract. *Postharvest Biol. Technol.* **2011**, *60*, 64–70. [CrossRef]

40. Fagundes, C.; Palou, L.; Monteiro, A.R.; Pérez-Gago, M.B. Hydroxypropyl methylcellulose-beeswax edible coatings formulated with antifungal food additives to reduce Alternaria black spot and maintain postharvest quality of cold-stored cherry tomatoes. *Sci. Hortic. (Amsterdam)* **2015**, *193*, 249–257. [CrossRef]

41. Fagundes, C.; Carciofi, B.A.M.; Monteiro, A.R. Estimate of respiration rate and physicochemical changes of fresh-cut apples stored under different temperatures. *Food Sci. Technol.* **2013**, *33*, 60–67. [CrossRef]

42. Ribeiro, C.; Vicente, A.A.; Teixeira, J.A.; Miranda, C. Optimization of edible coating composition to retard strawberry fruit senescence. *Postharvest Biol. Technol.* **2007**, *44*, 63–70. [CrossRef]

43. Harker, F.R.; Feng, J.; Johnston, J.W.; Gamble, J.; Alavi, M.; Hall, M.; Chheang, S.L. Influence of postharvest water loss on apple quality: The use of a sensory panel to verify destructive and non-destructive instrumental measurements of texture. *Postharvest Biol. Technol.* **2019**, *148*, 32–37. [CrossRef]

44. Tripathi, P.; Dubey, N.K.; Banerji, R.; Chansouria, J.P.N. Evaluation of some essential oils as botanical fungitoxicants in management of post-harvest rotting of citrus fruits. *World J. Microbiol. Biotechnol.* **2004**, *20*, 317–321. [CrossRef]

45. Basiak, E.; Linke, M.; Debeaufort, F.; Lenart, A.; Geyer, M. Dynamic behaviour of starch-based coatings on fruit surfaces. *Postharvest Biol. Technol.* **2019**, *147*, 166–173. [CrossRef]

46. Valencia-Chamorro, S.A.; Palou, L.; del Río, M.A.; Pérez-Gago, M.B. Antimicrobial edible films and coatings for fresh and minimally processed fruits and vegetables: A review. *Crit. Rev. Food Sci. Nutr.* **2011**, *51*, 872–900. [CrossRef] [PubMed]

47. Grande-Tovar, C.D.; Chaves-Lopez, C.; Serio, A.; Rossi, C.; Paparella, A. Chitosan coatings enriched with essential oils: Effects on fungi involve in fruit decay and mechanisms of action. *Trends Food Sci. Technol.* **2018**, *78*, 61–71. [CrossRef]

48. Campos-Requena, V.H.; Pérez, M.A.; Sanfuentes, E.A.; Figueroa, N.E.; Figueroa, C.R.; Rivas, B.L. Thermoplastic starch/clay nanocomposites loaded with essential oil constituents as packaging for strawberries—In vivo antimicrobial synergy over *Botrytis cinerea*. *Postharvest Biol. Technol.* **2017**, *129*, 29–36. [CrossRef]

49. Shao, X.; Cao, B.; Xu, F.; Xie, S.; Yu, D.; Wang, H. Effect of postharvest application of chitosan combined with clove oil against citrus green mold. *Postharvest Biol. Technol.* **2015**, *99*, 37–43. [CrossRef]

Article

Comparison of Properties of the Hybrid and Bilayer MWCNTs—Hydroxyapatite Coatings on Ti Alloy

Beata Majkowska-Marzec [1], Dorota Rogala-Wielgus [1], Michał Bartmański [1], Bartosz Bartosewicz [2] and Andrzej Zieliński [1,*]

1   Department of Materials Engineering and Bonding, Gdansk University of Technology, Narutowicza 11/12 str., 80-233 Gdansk, Poland; beamajko@pg.edu.pl (B.M.-M.); dorota.wielgus@pg.edu.pl (D.R.-W.); michal.bartmanski@pg.edu.pl (M.B.)

2   Institute of Optoelectronics, Military University of Technology, gen. Sylwestra Kaliskiego 2 str., 00-908 Warsaw, Poland; bartosz.bartosewicz@wat.edu.pl

*   Correspondence: andrzej.zielinski@pg.edu.pl; Tel.: +48-501-329-368; Fax: +48-583471815

Received: 16 September 2019; Accepted: 30 September 2019; Published: 4 October 2019

Abstract: Carbon nanotubes are proposed for reinforcement of the hydroxyapatite coatings to improve their adhesion, resistance to mechanical loads, biocompatibility, bioactivity, corrosion resistance, and antibacterial protection. So far, research has shown that all these properties are highly susceptible to the composition and microstructure of coatings. The present research is aimed at studies of multi-wall carbon nanotubes in three different combinations: multi-wall carbon nanotubes layer, bilayer coating composed of multi-wall carbon nanotubes deposited on nanohydroxyapatite deposit, and hybrid coating comprised of simultaneously deposited nanohydroxyapatite, multi-wall carbon nanotubes, nanosilver, and nanocopper. The electrophoretic deposition method was applied for the fabrication of the coatings. Atomic force microscopy, scanning electron microscopy and X-ray electron diffraction spectroscopy, and measurements of water contact angle were applied to study the chemical and phase composition, roughness, adhesion strength and wettability of the coatings. The results show that the pure multi-wall carbon nanotubes layer possesses the best adhesion strength, mechanical properties, and biocompatibility. Such behavior may be attributed to the applied deposition method, resulting in the high hardness of the coating and high adhesion of carbon nanotubes to the substrate. On the other hand, bilayer coating, and hybrid coating demonstrated insufficient properties, which could be the reason for the presence of soft porous hydroxyapatite and some agglomerates of nanometals in prepared coatings.

Keywords: hardness; adhesion; hydroxyapatite; carbon nanotubes; titanium; biomedical applications

---

1. Introduction

Carbon nanotube coatings (CNTs) demonstrate unique mechanical and biological properties. Thanks to this, they are increasingly applied in medicine and diagnostics, including tissue engineering [1,2]. These two-dimensional carbon structures are used, among others, to functionalize materials designed for implants, where CNTs can support osseointegration [3,4].

The biocompatibility of CNTs in orthopedic applications was established by in vitro studies, which showed accelerated bone growth and increased proliferation and differentiation of osteoblasts [5–10]. The most popular kind of metal substrate for CNTs is titanium [3,11–18], which combines some beneficial mechanical properties and biocompatibility with a chemical in vivo susceptibility [11]. Several studies evaluated the body's reaction in the presence of carbon nanotubes, demonstrating a high vitality of osteoblasts compared to the pure titanium substrate [14–16,19]. The ceramic coating consisting of multiple functionalized CNTs with carboxyl groups and hydroxyapatite (HAp) was reported to enhance mechanical properties and biological adhesion, as well as the response of osteoblasts [20–23]. CNTs have a unique chemical structure, so it could serve as a carrier, e.g., of an antibiotic [11] or other substances [21], able to prevent or cure potential infection in place of implantation and protect against implant rejection.

Nevertheless, possible toxicity of nanomaterials limits CNTs' medical applications [24], even though reports to date are scarce and inconclusive, e.g., for the neurotoxic effects of CNTs [25], or CNTs formation of reactive oxygen species [26]. There are more reports on the lack of adverse effects of CNTs than on their long-term toxicity [24]. Recent toxicological studies performed with the liver and kidney cells showed no adverse outcome [23]. In another study, neither SWCNTs (Single-Wall CNTs) nor MWCNTs (Mulit-Wall CNTs) demonstrated in vitro cytotoxicity for fibroblasts and hippocampal cells [27]. Moreover, CNTs are extensively investigated as components of biocoatings [2], e.g., for the Mg-phosphate coating reinforced by SiC nanowire-CNTs [28], HAp-CNTs composite coating on Mg alloy [29], SWCNTs/HAp and MWCNTs on Ti and its alloys [14–17]. Thus, the cytotoxicity of CNTs is generally assumed to be negligible. Even the use of both SWCNTs and MWCNTs as a base liquid with human blood was reported [30].

Pure SWCNTs, as well as MWCNTs coatings, have been obtained [31–33], but the composite or hybrid layers are even more extensively developed as, e.g., CNTs–HAp [13,22,23,34–37] or CNTs–graphene oxide [38]. Most frequently, the electrophoretic deposition (EPD) [12,14,39–41], electrocathodic deposition [15,16,20] or chemical vapor deposition (CVD) processes [42] are applied to prepare CNTs coating.

In the case of materials intended for dental or orthopedic load-bearing implants, adhesion and bond strength are essential, especially at the stage of implantation and ingrowth of human tissue. As Gopi et al. observed, the addition of low amounts of MWCNTs (0.5 and 1 mass pct.) increases hardness, and Young's modulus of the sol-gel derived HAp/MWCNTs coatings [15]. Nevertheless, the concentration of 1% and 2% of CNTs in the composite did not affect adhesion strength significantly and reached 24.2 and 22.4 MPa, respectively [16]. Mukherjee et al. reported an improvement of fracture toughness, flexural and impact strength values for MWCNTs-reinforced HAp [23]. The interfacial shear strength and the maximum load-bearing capacity of the tested CNTs–Ti interfaces were assessed at 37.8 MPa and 245 nN, respectively [43]. The tape tests displayed high adhesion strength (class 5B) for CNTs–(Zn)HAp coating [18]. In another work, the addition of CNTs to the Ti–HAp composite improved both the adhesion strength and hardness for the Ni–Ti substrate [35]. The highest bonding strength of 25.7 MPa was reported for the SWNTs/HAp coating and was nearly 70% more elevated than that of the pure HAp coating [17]. Better mechanical properties were observed for new complex hybrid materials, such as MWCNTs-HAp [20] and fluorohydroxyapatite (FA)-CNTs coating deposited on the TiO_2 nanotubular layer [44]. In the last case, CNTs served as a reinforcement, because of their higher elastic modulus compared to the FA matrix. The homogeneous distribution of decorated CNTs resulted in the robust interface between FA and CNTs. CNT-reinforced HAp composites had substantially better bending strength and fracture toughness than pure HAp [45]. Thus, the presence of MWCNTs evidently improves cohesion, but its adhesion to titanium substrate is less known and may be dependent on the specific architecture of a coating.

In sum, all the investigated material features such as biocompatibility, adhesion strength, and corrosion resistance are highly sensitive to coating composition and microstructure. The objective of the present research is to assess the adhesion strength, mechanical properties of coatings and wettability for the pure MWCNTs layer, hybrid CNTs–nanohydroxyapatite (nanoHAp) coating, and composite coating. Three different methods of deposition of carbon nanotubes were applied to obtain such coatings: (i) deposition of MWCNTs on a substrate surface, (ii) deposition of nanoHAp coating followed by the deposition of MWCNTs, and (iii) joint deposition of a mixture of nanoHAp, MWCNTs, and some nanometals. The purpose of such a choice lies in the fundamental importance of adhesion for the application of such surface treatment of load-bearing implants, because of the high stresses imposed on them during implantation surgery and the post-implantation period.

2. Materials and Methods

2.1. Preparation of Substrate Surfaces

The Ti13Nb13Zr alloy of the composition shown in Table 1 was used as a substrate. Specimens with a 40 mm diameter were cut from the rods. The surface was ground using abrasive paper SiC up to grit # 800. Then, the samples were rinsed with acetone, distilled water, air-dried, pickled in 5% HF for 30 s to remove oxide layers from the surface and finally rinsed with distilled water.

Table 1. Chemical composition of the Ti13Nb13Zr alloy.

Element	Nb	Zr	Fe	C	H	O	S	Hf	Ti
wt. pct.	13.18	13.49	0.085	0.035	0.004	0.078	<0.001	0.055	rem.

2.2. Preparation of CNTs' Suspension

To prepare the coatings, MWCNTs (3D-nano, number of walls 3–15, outer diameter 5–20 nm, inner diameter 2–6 nm, and length 1–10 μm) were functionalized in a mixture of concentrated sulfuric and nitric acid to add carboxyl groups and to provide a negative charge on the surface of carbon nanotubes. Four hundred and eighty grams of powder was annealed in a vacuum furnace (PROTHERM PC442, Ankara, Turkey) for 8 h at 400 °C and then dispersed in deionized water in an ultrasonic homogenizer (Bandelin Sonopuls HD 2070, Berlin, Germany). The suspension was added to 200 mL of mixed H_2SO_4 and at a ratio of 3:1 *v/v* and heated at 70 °C for 2 h [3]. To prepare the suspension of carbon nanotubes, the reaction mixture was centrifuged and washed several times with water or isopropanol until a neutral pH was reached. The concentration of CNTs in the obtained suspension was 0.27 wt.% in water and 0.4 wt.% in isopropanol. Final suspensions were sonicated for 1 min using an ultrasonic homogenizer to disperse the CNTs well after centrifugation. To prepare the mixed coating (m0.4CNT), 1.25 mL of 0.4 wt.% suspension of carbon nanotubes in isopropanol and 0.1 g of nanohydroxyapatite (grain size distribution approximately 20 nm, 99.8% purity, MKnano, Missisauga, Canada) were dispersed in 100 mL of ethyl alcohol (99.8% purity, Sigma Aldrich, St. Louis, MI, USA) and then mixed with 0.005 g of nanosilver (grain size distribution approx. 30 nm, Hongwu International Group Ltd., Guangzhou, China) and 0.005 g of nanocopper (grain size distribution approximately 80 nm, Hongwu International Group Ltd.) before carrying out the electrophoretic deposition (EPD) process.

2.3. Deposition of Coatings

The electrophoretic deposition (EPD) method was used to prepare the coatings. Their synthesis parameters are shown in Table 2. The Ti13Nb13Zr substrate was used as an anode and platinum as a counter electrode. The electrodes were placed parallel to each other at a distance of 5 mm and connected to a DC power source (MCP/SPN110-01C, Shanghai MCP Corp., Shanghai, China). The coatings were heated in a tubular furnace (PROTHERM PC442) from room temperature to 800 °C at a rate of 200 °C/h and cooled to room temperature with the oven.

Table 2. Parameters of synthesis of the coatings with multi-walled carbon nanotubes.

Coating	Synthesis Stage	Concentration of MWCNTs [%]	Duration of EPD [min]	EDP Voltage [V]	Temperature [°C]
0.27CNT	EPD of MWCNTs	0.27	2	11	ambient
H0.27CNT	EPD of nanoHAp	–	2	30	ambient
	Sintering	–	120	–	800
	EPD of MWCNTs	0.27	2	30	ambient
m0.4CNT	EPD of MWCNTs, nanoHAp, nanosilver, nanocopper	0.4	2	30	ambient
	Sintering	–	120	–	800

2.4. Structure and Morphology

An atomic force microscope (AFM NaniteAFM, Nanosurf, Bracknell, Great Britain) was used to study the surface topography. The examinations were performed in the non-contact mode at 20 mN force. The average roughness index S_a values were estimated based on 512 lines made in the area of $80.4 \times 80.4\ \mu m^2$.

The specimens' surfaces were observed using a high-resolution scanning electron microscope (SEM JEOL JSM-7800F, Tokyo, Japan) with a LED detector, at a 5 kV acceleration voltage.

The chemical composition of the coatings was investigated by an X-ray energy dispersive spectrometer (EDS Edax Inc., Mahwah, NJ, USA).

2.5. Nanomechanical Studies

Nanoindentation tests were performed with the NanoTest™ Vantage (Micro Materials, Wrexham, Great Britain) using a Berkovich three-sided pyramidal diamond. Twenty-five (5 × 5) measurements were carried out on each sample. The maximum applied force was 10 mN, the loading and unloading times were set up at 20 s and the dwell period at maximum load was 10 s. The distance between the subsequent indents was 20 μm. During the indent, the load–displacement curve was determined using the Oliver and Pharr method. Based on the load–penetration depth curves, the surface hardness (H) and Young's modulus (E) were calculated using integrated software. Estimating Young's modulus (E), the Poisson's ratio of 0.25 was assumed for carbon nanotube coatings and 0.36 for Ti13Nb13Zr.

Nanoscratch tests were performed with NanoTest™ Vantage (Micro Materials) using a Berkovich three-sided pyramidal diamond. The scratch tests were made by increasing the load from 0 to 200 mN at a loading rate of 1.3 mN/s at a distance of 500 μm. The adhesion of the coating was assessed based on the observation of an abrupt change in frictional force during the test.

2.6. Contact Angle Studies

Water contact angle measurements were carried out by falling drop method using a contact angle instrument (Contact Angle Goniometer, Zeiss, Oberkochen, Germany) at room temperature 10 s after drop out.

3. Results and Discussion

3.1. Structure and Morphology

Figure 1A,B show the AFM images of the topography of the native material, where substantial roughness of the surface after grinding and etching may be observed. After etching, the native material shows higher surface roughness.

Figure 1C–E illustrate the surface topography of the examined coatings composed of, respectively, carbon nanotubes (0.27CNT), carbon nanotubes deposited on the nanohydroxyapatite layer (H0.27CNT) and the composite of carbon nanotubes, nanohydroxyapatite, nanosilver and nanocopper (m0.4CNT). The addition of nanoHAp significantly increases roughness. Such a result may mean that the HAp penetrates through free spaces among nanotubes only a small amount. Fathyunes et al. (2018) proposed that during electrodeposition the water reduction produces hydroxyl ions, causing an increase in the pH near the cathode. Then, the CaP ceramics become insoluble and precipitate on the surface of the titanium substrate [46].

Figure 2 presents the SEM images of the surface topography of carbon nanotubes (0.27CNT), hybrid coating (H0.27CNT), and composite coating (m0.4CNT) together with their EDS spectra, confirming the formation of individual coatings. The carbon nanotubes can be distinguished for each specimen.

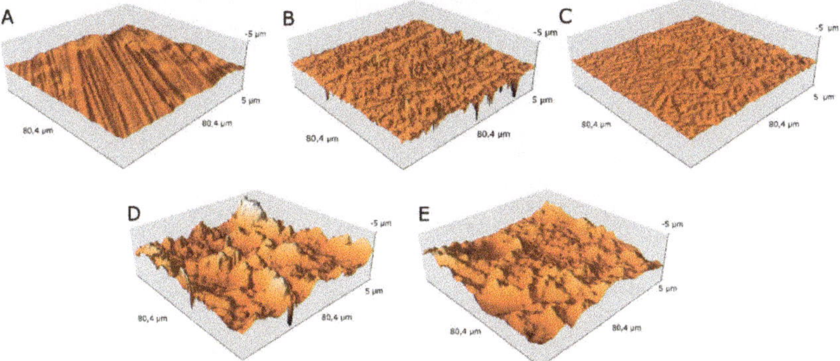

Figure 1. Atomic force microscope (AFM) surface topography of the: (**A**) reference sample—native material after grinding (MR), (**B**) reference sample—after etching (MRe), (**C**) sample of carbon nanotubes (0.27CNT), (**D**) sample of carbon nanotubes deposited on the nanohydroxyapatite layer(H0.27CNT), (**E**) sample of composite of carbon nanotubes, nanohydroxyapatite, nanosilver and nanocopper (m0.4CNT).

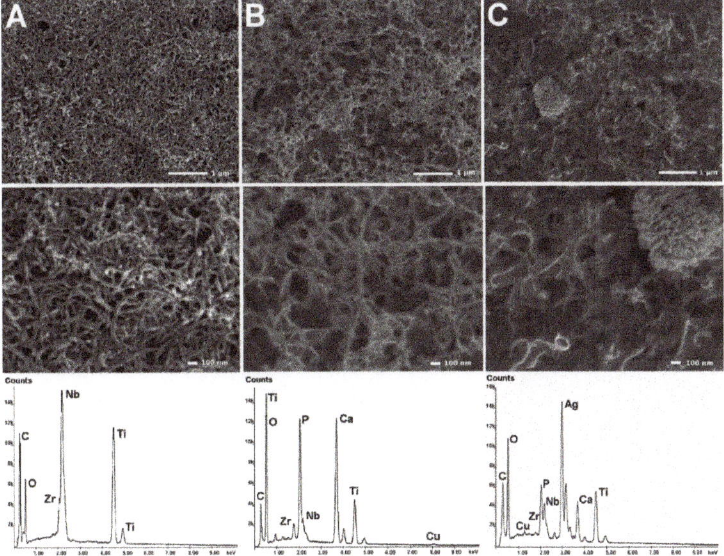

Figure 2. SEM surface topography with the energy dispersive spectrometer (EDS) spectrum of the sample: (**A**) 0.27CNT, (**B**) H0.27CNT, (**C**) m0.4CNT.

The SEM images (Figure 2) demonstrate a more uniform distribution of carbon nanotubes for the 0.27CNT sample than for the H0.27CNT sample. On the surface of the m0.4CNT sample, many agglomerates are present, as a result of the simultaneous deposition of nanosilver and nanocopper, which do not move into a bulk, but are absorbed on the HAp coating, resulting in decreased roughness (Table 3). The roughness values are similar to those previously reported [14].

Table 3. Surface roughness of the native material and deposited coatings.

Sample	Roughness S_a [μm]
MR	0.203
MRe	0.256
0.27CNT	0.098
H0.27CNT	0.980
M0.4CNT	0.618

3.2. Mechanical Studies

3.2.1. Nanoindentation

Figure 3 shows the load–displacement hysteresis curve as an example of the results of nanoindentation tests. Three stages of the nanoindentation test are observed: the raising load to a maximum value, the pause (to stabilize the probe at maximum depth), and offloading. An irregularity in the form of a step due to temperature drift, adjusted at the end of nanoindentation, can be observed.

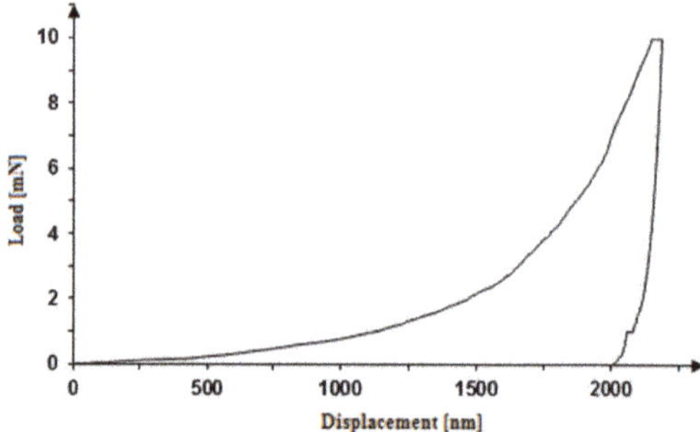

Figure 3. Nanoindentation load–displacement curve obtained for pure CNT coating (0.27CNT).

The exact values of measured mechanical properties are listed in Table 4. The lowest hardness values are demonstrated for H0.26CNT and m0.4CNT coatings due to the significant effect of etching on hardness. A 53-fold difference in the hardness is noticed between polished and etched specimens. The CNTs alone have the best hardness and wear resistance. The addition of softer nanoHAp decreases mechanical strength. The composite coating displays improved behavior, presumably because of a different coating architecture and the decisive role of metal nanoparticles. The hardness here measured is lower than the values previously reported in [44] as 0.37 to 0.58 GPa.

The elastic modulus values observed here are in line with some previous results for titanium, and much smaller for CNTs coating, so far reported as 60 MPa [14] and 113–130 MPa [16]. An increase in Young's modulus from 15 to 40 GPa [35] and from 12 to 19 GPa [44] was also observed. The reason for such discrepancies may be the high dependence of nanomechanical properties of the coating architecture, test parameters, and fractions of components.

Table 4. Mechanical properties and maximum indent depth for the substrate and achieved coatings.

Sample	Nanohardness [GPa]	Reduced Young's Modulus [GPa]	Young's Modulus [GPa]	Maximum Indent Depth [nm]
I	3.758 ± 1.045	116.91 ± 16.32	83.32 ± 11.63	330.92 ± 36.61
MRe	0.071 ± 0.010	14.57 ± 2.21	9.44 ± 1.43	2358.11 ± 170.08
0.27CNT	0.101 ± 0.049	18.59 ± 5.66	14.17 ± 4.32	2069.67 ± 352.57
H0.27CNT	0.022 ± 0.015	7.46 ± 3.65	5.63 ± 2.76	4264.18 ± 1150.11
m0.4CNT	0.035 ± 0.019	11.72 ± 4.31	8.88 ± 3.26	3210.02 ± 817.53

Figures 4 and 5 show the 3D distribution of Young's modulus and nanohardness for the examined samples. Compared to the other materials, the native material after grinding (MR) reveals the biggest nanohardness and Young modulus.

The 3D Young's modulus (Figure 4) and nanohardness (Figure 5) distribution graphs show a highly non-uniform rough surface. The roughest is only a polished surface, lower roughness is observed for CNTs coating, and the etched surface and two other coatings show differences in Young's modulus in the range of 20 MPa. For hardness, the same effects can be noticed. The most heterogeneous are the H0.27CNT and m0.4CNT coatings, due either to appearing agglomerates or variable layer thickness, or both. The 0.27CNT specimen's nanohardness and Young's modulus distribution graphs show "uplifts," which could result from the probe's contact with the surface of the native material. Nevertheless, the 0.27CNT sample appears to possess a Young's modulus very similar to that of a human bone. Cuppone et al. reported that the cortical bone has an average Young's modulus value of 18.6 ± 1.9 GPa [47]. In conclusion, the results generally show that Young's modulus increases with rising nanohardness.

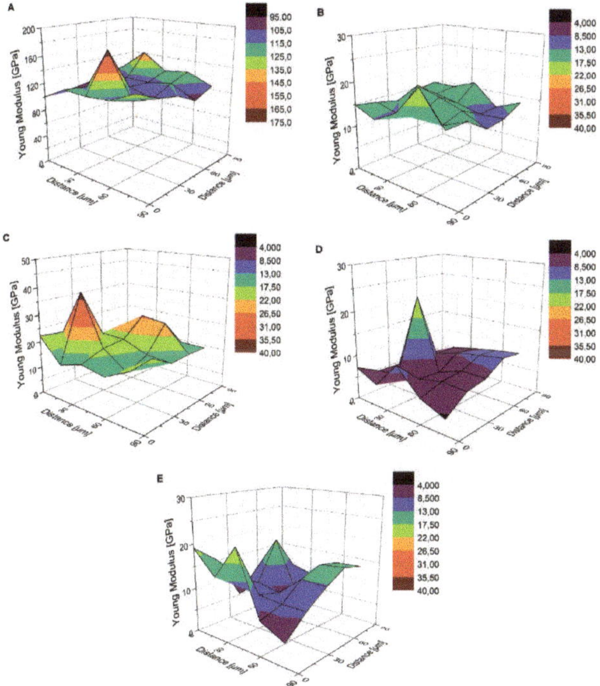

Figure 4. 3D Young's modulus distribution for: (**A**) native material after grinding (MR), (**B**) native material after etching (MRe), (**C**) CNTs coating (0.27CNT), (**D**) CNTs deposited on HAp coating (H0.27CNT), (**E**) mixed coating consisting of CNTs, nanoHAp, nanoAg and nanoCu (m0.4CNT).

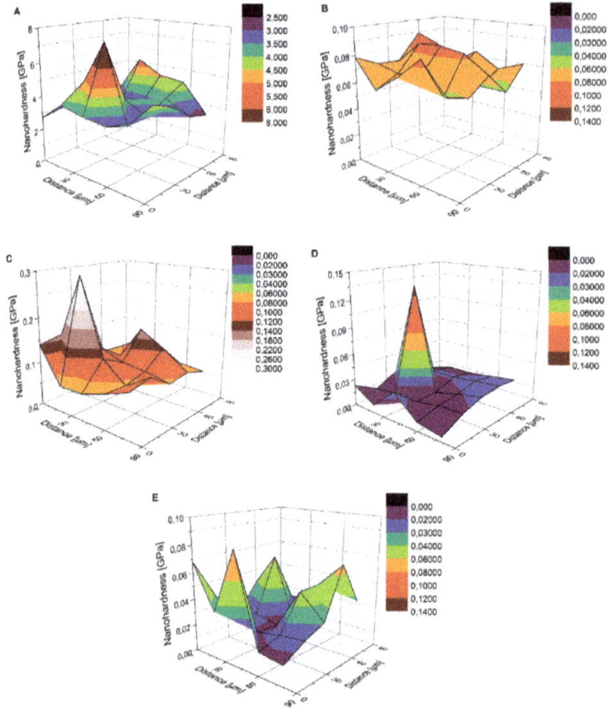

Figure 5. 3D nanohardness distribution for (**A**) native material after grinding (MR), (**B**) native material after etching (MRe), (**C**) CNTs coating (0.27CNT), (**D**) CNTs deposited on HAp coating (H0.27CNT), (**E**) mixed coating consisting of CNTs, nanHAp, nanoAg and nanoCu (m0.4CNT).

3.2.2. Nanoscratch Test

Figure 6 shows the relation of friction (F) on load (L) for each coating subjected to the nanoscratch test. The graphs describe a critical load (Lc) for every single measurement, represented by a vertical line indicating the moment of delamination for the carbon nanotube coating (0.27CNT), the carbon nanotube coating on hydroxyapatite (H0.27CNT) and the mixed coating consisting of carbon nanotubes, nanohydroxyapatite, nanosilver and nanocopper (m0.4CNT), respectively.

The values of the critical load (Lc) and critical friction (Fc), which indicate the load and friction under which the coating cracks or is delaminated, are shown in Table 5.

To conclude, the CNTs coating deposited on the Ti13Nb13Zr alloy (0.27CNT) has the best strength adhesion to the surface, while the worst adhesion is demonstrated by the composite coating (m0.4CNT). Application of HAp ceramics as an interlayer and its sintering does not improve the adhesion of the CNTs coating to the surface of titanium alloy. What is more, the addition of nanosilver and nanocopper to the composite coating further decreases adhesion, presumably due to the change in the particle size of nanometals caused by agglomeration in the bath. These results demonstrate that adhesion is best when the CNTs adhere directly to the surface, forming strong chemical bonds. Regretfully, these results cannot be compared to the adhesion strength 18–22 MPa measured by the Adhesion Test method [16] and 32 MPa measured by the F1044 shear bond strength test [35] as these methods are very different from the nanoscratch tests. On the other hand, during surgery and the period after implantation, the coatings are subject to shear stresses. Therefore, the nanoscratch method seems particularly suitable for determining real adhesion.

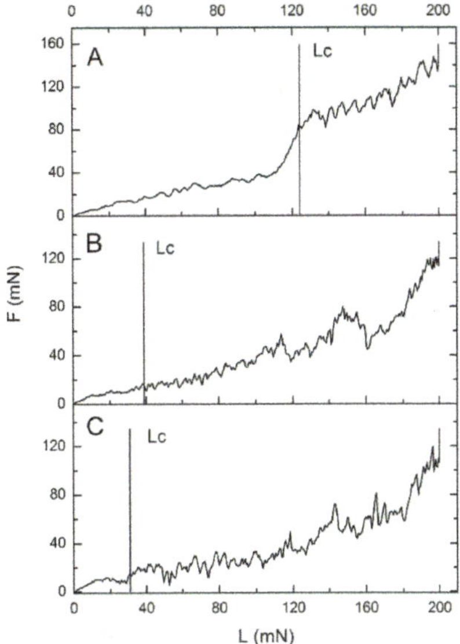

Figure 6. Friction (F) dependence on load (L) with the indicated critical load (Lc) for the single measurement of (**A**) the carbon nanotube coating on the surface of Ti13Nb13Zr (0.27CNT), (**B**) the carbon nanotube coating on the surface of the hydroxyapatite (H0.27CNT), (**C**) the mixed coating consisting of carbon nanotubes, nanohydroxyapatite, nanosilver and nanocopper (m0.4CNT).

Table 5. Parameters of coatings delamination.

Sample	Critical Friction (Fc) [mN]	Critical Load (Lc) [mN]
0.27CNT	89.42 ± 36.19	116.50 ± 32.07
H0.27CNT	39.96 ± 18.07	92.06 ± 34.3
m0.4CNT	29.56 ± 6.92	60.38 ± 10.21

3.3. Contact Angle Measurements

Figure 7 shows images of the water drops poured on the surface of all tested specimens. After etching, the surface of the Ti13Nb13Zr became hydrophobic, as the mean contact angle reached the value of 97.40°. This could be the result of the higher roughness of the etched surface (Table 3). The CNTs coating was found hydrophilic [48] with the angle value 56.50°, which means that the surface is proper for its application in implantology. Türka et al. achieved similar results for the contact angle of the CNTs after functionalization, 53.29° [49]. However, the CNTs–HAp and composite coatings are both hydrophobic, and as such they cannot be useful for implants. Prodana et al. explained the hydrophilicity of TiO_2/MWCNTs/HAp coating on titanium substrate by the appearance of C–O single bonds, and C=O and O–C=O more oxidized forms, which might be responsible for increasing surface hydrophilicity [20]. Nevertheless, during the electrophoretic deposition process, the pH rises at the cathodic substrate, resulting in deprotonation of carboxyl groups (COOH) and, thereby, functionalized CNTs become more negatively charged. Carboxyl groups (COO–) facilitate the electrostatic interaction of Ca^{2+} ions from HAp with CNTs [46]. Based on reports to date, and our research, the hydrophobic appearance of an arrangement of CNTs and porous HAp may be attributed to an occurrence of a specific architecture with a lesser ability to form van der Waals bonds between water and specimen surface. A similar explanation may be given regarding the composite materials, indicating the increased effects of nanohydroxyapatite, which may have a different microstructure to CNTs. Further attempts made with various forms of both components, oriented particularly towards better hydrophilicity, should give a more plausible solution.

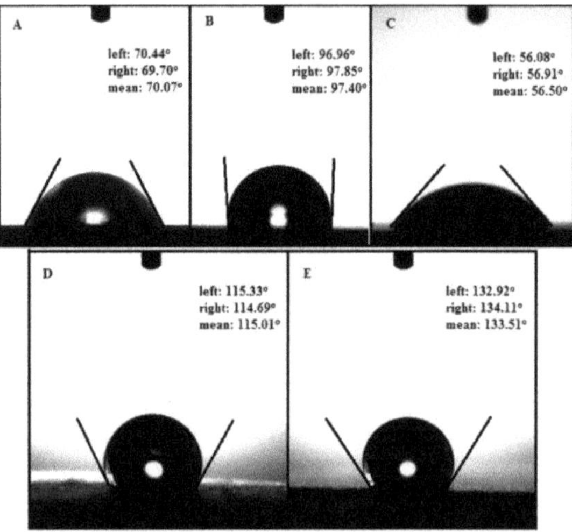

Figure 7. Contact angle (CA) for: (**A**) native material after grinding (MR), (**B**) native material after etching (MRe), (**C**) carbon nanotube coating on the surface of Ti13Nb13Zr (0.27CNT), (**D**) carbon nanotube coating on the surface of the hydroxyapatite (H0.27CNT), (**E**) the mixed coating consisting of carbon nanotubes, nanohydroxyapatite, nanosilver and nanocopper (m0.4CNT).

4. Conclusions

The multi-wall carbon nanotubes deposited on the surface of the Ti13Nb13Zr alloy by the electrophoretic method demonstrate a relatively high hardness and wear resistance, firm adhesion, and proper wettability, so they are suitable for surface treatment of the titanium implants made of the investigated alloy.

The hybrid coatings obtained by the formation of hydroxyapatite deposit, followed by a carbon nanotubes' layer, have lower adhesion strength, hardness, and less proper wettability, presumably due to the weak interface bonding between hydroxyapatite and carbon nanotubes, which is related to the properties of the ceramic material.

The composite layer has the lowest adhesion strength, hardness, and equally improper wettability, which can be attributed to weak bonding of components in the coating.

The obtained results provide evidence that the mutual bonding strength of all components and critical strength at the coating–substrate interface, are more significant determinants of mechanical properties than the components and the deposition process parameters.

Author Contributions: Conceptualization, B.M.-M.; Formal Analysis, M.B. and B.B.; Investigation, B.M.-M., D.R.-W. and M.B.; Methodology, B.M.-M., D.R.-W., M.B. and B.B.; Project administration, B.M.-M.; Resources, B.B.; Supervision, A.Z.; Visualization, D.R.-W.; Writing–Original Draft, B.M.-M. and D.R.-W.; Writing–Review and Editing, A.Z.

Funding: This research received no external funding.

Acknowledgments: We are grateful to Julia Mazurowska for her help in carrying out experiments and Bartłomiej Jankiewicz's team from the Military University of Technology, Institute of Optoelectronics for sharing laboratory equipment and support in preparing materials to tests.

Conflicts of Interest: The authors declare no conflict of interest.

References

1. Ku, S.H.; Lee, M.; Park, C.B. Carbon-Based Nanomaterials for Tissue Engineering. *Adv. Healthc. Mater.* **2013**, *2*, 244–260. [CrossRef]
2. Li, X.; Liu, X.; Huang, J.; Fan, Y.; Cui, F. Surface & Coatings Technology Biomedical investigation of CNT based coatings. *Surf. Coat. Technol.* **2011**, *206*, 759–766.
3. Dlugon, E.; Simka, W.; Fraczek-Szczypta, A.; Niemiec, W.; Markowski, J.; Szymanska, M.; Blazewicz, M. Carbon nanotube-based coatings on titanium. *Bull. Mater. Sci.* **2015**, *38*, 1339–1344. [CrossRef]
4. Tanaka, M.; Sato, Y.; Zhang, M.; Haniu, H.; Okamoto, M.; Aoki, K.; Takizawa, T.; Yoshida, K.; Sobajima, A.; Kamanaka, T.; et al. In Vitro and In Vivo Evaluation of a Three-Dimensional Porous Multi-Walled Carbon Nanotube Scaffold for Bone Regeneration. *Nanomaterials* **2017**, *7*, 46. [CrossRef] [PubMed]
5. Lahiri, D.; Ghosh, S.; Agarwal, A. Carbon nanotube reinforced hydroxyapatite composite for orthopedic application: A review. *Mater. Sci. Eng. C* **2012**, *32*, 1727–1758. [CrossRef]
6. Usui, Y.; Aoki, K.; Narita, N.; Murakami, N.; Nakamura, I.; Nakamura, K.; Ishigaki, N.; Yamazaki, H.; Horiuchi, H.; Kato, H.; et al. full papers Carbon Nanotubes with High Bone-Tissue Compatibility and Bone-Formation Acceleration Effects. *Small* **2008**, *8621*, 240–246. [CrossRef] [PubMed]
7. Kalbacova, M.; Kalbac, M. Influence of single-walled carbon nanotube films on metabolic activity and adherence of human osteoblasts. *Carbon* **2007**, *45*, 2266–2272. [CrossRef]
8. Lahiri, D.; Benaduce, A.P.; Rouzaud, F.; Solomon, J.; Keshri, A.K.; Kos, L.; Agarwal, A. Wear behavior and in vitro cytotoxicity of wear debris generated from hydroxyapatite–carbon nanotube composite coating. *J. Biomed. Mater. Res. Part A* **2010**, *0494*, 1–12. [CrossRef]
9. Matsuoka, M.; Akasaka, T.; Totsuka, Y.; Watari, F. Strong adhesion of Saos-2 cells to multi-walled carbon nanotubes. *Mater. Sci. Eng. B* **2010**, *173*, 182–186. [CrossRef]
10. Akasaka, T.; Yokoyama, A.; Matsuoka, M.; Hashimoto, T.; Watari, F. Thin fi lms of single-walled carbon nanotubes promote human osteoblastic cells (Saos-2) proliferation in low serum concentrations. *Mater. Sci. Eng. C* **2010**, *30*, 391–399. [CrossRef]
11. Hirschfeld, J.; Akinoglu, E.M.; Wirtz, D.C.; Hoerauf, A.; Bekeredjian-Ding, I.; Jepsen, S.; Haddouti, E.M.; Limmer, A.; Giersig, M. Long-term release of antibiotics by carbon nanotube-coated titanium alloy surfaces diminish biofilm formation by Staphylococcus epidermidis. *Nanomed. Nanotechnol. Biol. Med.* **2017**, *13*, 1587–1593. [CrossRef] [PubMed]
12. Bai, Y.; Prasad, M.; Song, I.; Ho, M.; Sung, T.; Watari, F.; Uo, M. Electrophoretic deposition of carbon nanotubes – hydroxyapatite nanocomposites on titanium substrate. *Mater. Sci. Eng. C* **2010**, *30*, 1043–1049. [CrossRef]
13. Długoń, E.; Niemiec, W.; Fraczek-Szczypta, A.; Jeleń, P.; Sitarz, M.; Błazewicz, M. Spectroscopic studies of electrophoretically deposited hybrid HAp/CNT coatings on titanium. *Spectrochim. Acta Part A Mol. Biomol. Spectrosc.* **2014**, *133*, 872–875. [CrossRef]
14. Abrishamchian, A.; Hooshmand, T.; Mohammadi, M.; Najafi, F. Preparation and characterization of multi-walled carbon nanotube/hydroxyapatite nanocomposite film dip coated on Ti-6Al-4V by sol-gel method for biomedical applications: An in vitro study. *Mater. Sci. Eng. C* **2013**, *33*, 2002–2010. [CrossRef]
15. Gopi, D.; Shinyjoy, E.; Kavitha, L. Influence of ionic substitution in improving the biological property of carbon nanotubes reinforced hydroxyapatite composite coating on titanium for orthopedic applications. *Ceram. Int.* **2015**, *41*, 5454–5463. [CrossRef]
16. Gopi, D.; Shinyjoy, E.; Sekar, M.; Surendiran, M.; Kavitha, L.; Sampath Kumar, T.S. Development of carbon nanotubes reinforced hydroxyapatite composite coatings on titanium by electrodeposition method. *Corros. Sci.* **2013**, *73*, 321–330. [CrossRef]
17. Pei, X.; Zeng, Y.; He, R.; Li, Z.; Tian, L.; Wang, J.; Wan, Q.; Li, X.; Bao, H. Single-walled carbon nanotubes/hydroxyapatite coatings on titanium obtained by electrochemical deposition. *Appl. Surf. Sci.* **2014**, *295*, 71–80. [CrossRef]
18. Zhong, Z.; Qin, J.; Ma, J. Electrophoretic deposition of biomimetic zinc substituted hydroxyapatite coatings with chitosan and carbon nanotubes on titanium. *Ceram. Int.* **2015**, *41*, 8878–8884. [CrossRef]
19. Zanello, L.P.; Zhao, B.; Hu, H.; Haddon, R.C. Bone cell proliferation on carbon nanotubes. *Nano Lett.* **2006**, *6*, 562–567. [CrossRef]

20. Prodana, M.; Duta, M.; Ionita, D.; Bojin, D.; Stan, M.S.; Dinischiotu, A.; Demetrescu, I. A new complex ceramic coating with carbon nanotubes, hydroxyapatite and TiO 2 nanotubes on Ti surface for biomedical applications. *Ceram. Int.* **2015**, *41*, 6318–6325. [CrossRef]

21. Chouirfa, H.; Bouloussa, H.; Migonney, V.; Falentin-Daudré, C. Review of titanium surface modification techniques and coatings for antibacterial applications. *Acta Biomater.* **2019**, *83*, 37–54. [CrossRef] [PubMed]

22. Sivaraj, D.; Vijayalakshmi, K. Substantial effect of magnesium incorporation on hydroxyapatite/carbon nanotubes coatings on metallic implant surfaces for better anticorrosive protection and antibacterial ability. *J. Anal. Appl. Pyrolysis* **2018**, *135*, 15–21. [CrossRef]

23. Mukherjee, S.; Nandi, S.K.; Kundu, B.; Chanda, A.; Sen, S.; Das, P.K. Enhanced bone regeneration with carbon nanotube reinforced hydroxyapatite in animal model. *J. Mech. Behav. Biomed. Mater.* **2016**, *60*, 243–255. [CrossRef] [PubMed]

24. Malik, M.A.; Wani, M.Y.; Hashim, M.A.; Nabi, F. Nanotoxicity: Dimensional and morphological concerns. *Adv. Phys. Chem.* **2011**, *2011*.

25. Teleanu, D.; Chircov, C.; Grumezescu, A.; Teleanu, R. Neurotoxicity of Nanomaterials: An Up-to-Date Overview. *Nanomaterials* **2019**, *9*, 96. [CrossRef]

26. Mohanta, D.; Patnaik, S.; Sood, S.; Das, N. Carbon nanotubes: Evaluation of toxicity at biointerfaces. *J. Pharm. Anal.* **2019**. [CrossRef]

27. Nawrotek, K.; Tylman, M.; Rudnicka, K.; Gatkowska, J.; Balcerzak, J. Tubular electrodeposition of chitosan-carbon nanotube implants enriched with calcium ions. *J. Mech. Behav. Biomed. Mater.* **2016**, *60*, 256–266. [CrossRef]

28. Guan, K.; Zhang, L.; Zhu, F.; Sheng, H.; Li, H. Surface modification for carbon/carbon composites with Mg-CaP coating reinforced by SiC nanowire-carbon nanotube hybrid for biological application. *Appl. Surf. Sci.* **2019**, *489*, 856–866. [CrossRef]

29. Khazeni, D.; Saremi, M.; Soltani, R. Development of HA-CNTs composite coating on AZ31 Magnesium alloy by cathodic electrodeposition. Part 2: Electrochemical and in-vitro behavior. *Ceram. Int.* **2019**, *45*, 11186–11194. [CrossRef]

30. Alsagri, A.S.; Nasir, S.; Gul, T.; Islam, S.; Nisar, K.S.; Shah, Z.; Khan, I. MHD thin film flow and thermal analysis of blood with CNTs nanofluid. *Coatings* **2019**, *9*, 175. [CrossRef]

31. Przekora, A.; Benko, A.; Nocun, M.; Wyrwa, J.; Blazewicz, M.; Ginalska, G. Titanium coated with functionalized carbon nanotubes—A promising novel material for biomedical application as an implantable orthopaedic electronic device. *Mater. Sci. Eng. C* **2014**, *45*, 287–296. [CrossRef] [PubMed]

32. Jacobs, C.B.; Peairs, M.J.; Venton, B.J. Analytica Chimica Acta Review: Carbon nanotube based electrochemical sensors for biomolecules. *Anal. Chim. Acta* **2010**, *662*, 105–127. [CrossRef] [PubMed]

33. Benko, A.; Nocuń, M.; Berent, K.; Gajewska, M.; Klita, Ł.; Wyrwa, J.; Błażewicz, M. Diluent changes the physicochemical and electrochemical properties of the electrophoretically-deposited layers of carbon nanotubes. *Appl. Surf. Sci.* **2017**, *403*, 206–217. [CrossRef]

34. Sivaraj, D.; Vijayalakshmi, K. Novel synthesis of bioactive hydroxyapatite/f-multiwalled carbon nanotube composite coating on 316L SS implant for substantial corrosion resistance and antibacterial activity. *J. Alloys Compd.* **2019**, 1340–1346. [CrossRef]

35. Maleki-Ghaleh, H.; Khalil-Allafi, J. Characterization, mechanical and in vitro biological behavior of hydroxyapatite-titanium-carbon nanotube composite coatings deposited on NiTi alloy by electrophoretic deposition. *Surf. Coatings Technol.* **2019**, *363*, 179–190. [CrossRef]

36. Mohajernia, S.; Pour-Ali, S.; Hejazi, S.; Saremi, M.; Kiani-Rashid, A.R. Hydroxyapatite coating containing multi-walled carbon nanotubes on AZ31 magnesium: Mechanical-electrochemical degradation in a physiological environment. *Ceram. Int.* **2018**, *44*, 8297–8305. [CrossRef]

37. Park, J.E.; Jang, Y.S.; Bae, T.S.; Lee, M.H. Multi-walled carbon nanotube coating on alkali treated TiO2 nanotubes surface for improvement of biocompatibility. *Coatings* **2018**, *8*, 159. [CrossRef]

38. Fraczek-szczypta, A.; Jantas, D.; Ciepiela, F.; Grzonka, J.; Bernasik, A.; Marzec, M. Diamond & Related Materials Carbon nanomaterials coatings – Properties and in fl uence on nerve cells response. *Diam. Relat. Mater.* **2018**, *84*, 127–140.

39. Farrokhi-Rad, M.; Menon, M. Effect of Dispersants on the Electrophoretic Deposition of Hydroxyapatite-Carbon Nanotubes Nanocomposite Coatings. *J. Am. Ceram. Soc.* **2016**, *99*, 2947–2955. [CrossRef]

40. Liu, S.; Li, H.; Su, Y.; Guo, Q.; Zhang, L. Preparation and properties of in-situ growth of carbon nanotubes reinforced hydroxyapatite coating for carbon/carbon composites. *Mater. Sci. Eng. C* **2017**, *70*, 805–811. [CrossRef]
41. Singh, I.; Kaya, C.; Shaffer, M.S.; Thomas, B.C.; Boccaccini, A.R. Bioactive ceramic coatings containing carbon nanotubes on metallic substrates by electrophoretic deposition. *J. Mater. Sci.* **2006**, *41*, 8144–8151. [CrossRef]
42. Constanda, S.; Stan, M.S.; Ciobanu, C.S.; Motelica-Heino, M.; Guégan, R.; Lafdi, K.; Dinischiotu, A.; Predoi, D. Carbon Nanotubes-Hydroxyapatite Nanocomposites for an Improved Osteoblast Cell Response. *J. Nanomater.* **2016**, *2016*. [CrossRef]
43. Yi, C.; Bagchi, S.; Dmuchowski, C.M.; Gou, F.; Chen, X.; Park, C.; Chew, H.B.; Ke, C. Direct nanomechanical characterization of carbon nanotube - titanium interfaces. *Carbon N. Y.* **2018**, *132*, 548–555. [CrossRef]
44. Sasani, N.; Vahdati Khaki, J.; Mojtaba Zebarjad, S. Characterization and nanomechanical properties of novel dental implant coatings containing copper decorated-carbon nanotubes. *J. Mech. Behav. Biomed. Mater.* **2014**, *37*, 125–132. [CrossRef]
45. Zhao, X.; Chen, X.; Zhang, L.; Liu, Q.; Wang, Y.; Zhang, W.; Zheng, J. Preparation of Nano-Hydroxyapatite Coated Carbon Nanotube Reinforced Hydroxyapatite Composites. *Coatings* **2018**, *8*, 357. [CrossRef]
46. Fathyunes, L.; Khalil-Allafi, J.; Moosavifar, M. Development of graphene oxide/calcium phosphate coating by pulse electrodeposition on anodized titanium: Biocorrosion and mechanical behavior. *J. Mech. Behav. Biomed. Mater.* **2019**, *90*, 575–586. [CrossRef]
47. Cuppone, M.; Seedhom, B.B.; Berry, E.; Ostell, A.E. The Longitudinal Young's Modulus of Cortical Bone in the Midshaft of Human Femur and its Correlation with CT Scanning Data. *Calcif. Tissue Int.* **2004**, *74*, 302–309.
48. Sansotera, M.; Talaeemashhadi, S.; Gambarotti, C.; Pirola, C.; Longhi, M.; Ortenzi, M.A.; Navarrini, W.; Bianchi, C.L. Comparison of branched and linear perfluoropolyether chains functionalization on hydrophobic, morphological and conductive properties of multi-walled carbon nanotubes. *Nanomaterials* **2018**, *8*, 176. [CrossRef]
49. Türk, S.; Altınsoy, I.; Çelebi Efe, G.; Ipek, M.; Özacar, M.; Bindal, C. 3D porous collagen/functionalized multiwalled carbon nanotube/chitosan/hydroxyapatite composite scaffolds for bone tissue engineering. *Mater. Sci. Eng. C* **2018**, *92*, 757–768. [CrossRef]

 *coatings*

Article

Fibre Laser Treatment of Beta TNZT Titanium Alloys for Load-Bearing Implant Applications: Effects of Surface Physical and Chemical Features on Mesenchymal Stem Cell Response and *Staphylococcus aureus* Bacterial Attachment

Clare Lubov Donaghy [1,*], **Ryan McFadden** [1], **Graham C. Smith** [2], **Sophia Kelaini** [3], **Louise Carson** [4], **Savko Malinov** [1], **Andriana Margariti** [3] and **Chi-Wai Chan** [1,*]

[1] School of Mechanical and Aerospace Engineering, Queen's University Belfast, Belfast BT9 5AH, UK; rmcfadden02@qub.ac.uk (R.M.); S.Malinov@qub.ac.uk (S.M.)

[2] Department of Natural Sciences, University of Chester, Thornton Science Park, Chester CH2 4NU, UK; graham.smith@chester.ac.uk

[3] Centre for Experimental Medicine, Queen's University Belfast, Belfast BT9 7BL, UK; s.kelaini@qub.ac.uk (S.K.); a.margariti@qub.ac.uk (A.M.)

[4] School of Pharmacy, Queen's University Belfast, Belfast BT9 7BL, UK; l.carson@qub.ac.uk

* Correspondence: cdonaghy38@qub.ac.uk (C.L.D.); c.w.chan@qub.ac.uk (C.-W.C.)

Received: 25 February 2019; Accepted: 8 March 2019; Published: 12 March 2019

Abstract: A mismatch in bone and implant elastic modulus can lead to aseptic loosening and ultimately implant failure. Selective elemental composition of titanium (Ti) alloys coupled with surface treatment can be used to improve osseointegration and reduce bacterial adhesion. The biocompatibility and antibacterial properties of Ti-35Nb-7Zr-6Ta (TNZT) using fibre laser surface treatment were assessed in this work, due to its excellent material properties (low Young's modulus and non-toxicity) and the promising attributes of fibre laser treatment (very fast, non-contact, clean and only causes changes in surface without altering the bulk composition/microstructure). The TNZT surfaces in this study were treated in a high speed regime, specifically 100 and 200 mm/s, (or 6 and 12 m/min). Surface roughness and topography (WLI and SEM), chemical composition (SEM-EDX), microstructure (XRD) and chemistry (XPS) were investigated. The biocompatibility of the laser treated surfaces was evaluated using mesenchymal stem cells (MSCs) cultured in vitro at various time points to assess cell attachment (6, 24 and 48 h), proliferation (3, 7 and 14 days) and differentiation (7, 14 and 21 days). Antibacterial performance was also evaluated using *Staphylococcus aureus* (*S. aureus*) and Live/Dead staining. Sample groups included untreated base metal (BM), laser treated at 100 mm/s (LT100) and 200 mm/s (LT200). The results demonstrated that laser surface treatment creates a rougher (Ra value of BM is 199 nm, LT100 is 256 nm and LT200 is 232 nm), spiky surface (Rsk > 0 and Rku > 3) with homogenous elemental distribution and decreasing peak-to-peak distance between ripples (0.63 to 0.315 µm) as the scanning speed increases (p < 0.05), generating a surface with distinct micron and nano scale features. The improvement in cell spreading, formation of bone-like nodules (only seen on the laser treated samples) and subsequent four-fold reduction in bacterial attachment (p < 0.001) can be attributed to the features created through fibre laser treatment, making it an excellent choice for load bearing implant applications. Last but not least, the presence of TiN in the outermost surface oxide might also account for the improved biocompatibility and antibacterial performances of TNZT.

Keywords: mesenchymal stem cell (MSC); antibacterial performance; TNZT; beta titanium; fibre laser treatment

---

1. Introduction

Aseptic loosening is the most commonly cited indication of load bearing orthopaedic implant revision surgeries [1]. A mismatch in elastic modulus of bone and implant means that required stresses for bone remodelling are not obtained, leading to stress shielding which causes bone resorption and ultimately results in aseptic loosening. The ability of an implant to successfully integrate into native tissue is determined by the surface features, namely surface roughness and topography (physical), wettability (physiochemical) and chemistry (chemical). All of these play a crucial role in modulating cell–surface interactions. The ideal surface conditions for optimal osseointegration still remain to be fully elucidated, however, a consensus exists that surface features are vital in the control of cell response (adhesion, proliferation and differentiation). Among several cell types involved in the osseointegration process, mesenchymal stem cells (MSCs) are multipotent progenitor cells that are responsible for self-replicating and differentiating into varying lineages. The osteogenic lineage includes two fundamental cell types, osteocytes and osteoblasts, which are responsible for the formation of new bone. The hematopoietic-derived osteoclasts resorb bone cells and are crucial for bone homeostasis [2]. Insufficient mechanical loads at the implant site lead to exacerbated osteoclast activity [3], which can potentially be reduced by choosing a material with lower elastic modulus, a property belonging to beta (β) titanium (Ti) alloys.

Ti-based alloys, namely commercially pure (cp) Ti and Ti-6Al-4V, have been successfully applied for orthopaedic applications in the past few decades on account of their promising mechanical and corrosion properties, as well as desirable biocompatibility. In principle, they are classified as alpha (α), near-α, $\alpha + \beta$, metastable β or stable β, depending upon material composition (α or β-stabilizers) and thermo-mechanical processing history. β-stabilizers, such as Nb and Ta, are isomorphous, while Zr is a neutral stabilizer [4]. β Ti alloys offer unique characteristics in comparison with their cp Ti (α Ti) and Ti-6Al-4V ($\alpha + \beta$ Ti) counterparts. Among a family of β Ti alloys which have been developed for orthopaedic applications in recent years, the Ti-Nb-Zr-Ta (TNZT) quaternary alloy system is particularly promising due to its excellent properties, which include superior biocompatibility, low elastic modulus (55 GPa) [5], good corrosion resistance and the absence of toxic elements such as aluminium (Al) and vanadium (V). The cytotoxicity and adverse tissue reaction caused by Al and V have been widely reported in literature [6–9]. Another notable flaw in the choice of the conventional material used for load-bearing orthopaedic implants is the staggering difference in elastic modulus of cortical bone (7–30 GPa) [10] and Ti-6Al-4V (110 GPa) [11]. The utilisation of TNZT can aid in reducing aseptic loosening and stress shielding, and subsequently can cause a reduction in tissue reaction to particulate debris, while surface modification in the form of fibre laser treatment can be implemented to reduce bacterial infection and improve native cell adhesion.

Bacterial infection caused by *Staphylococcus aureus* (*S. aureus*) is considered to be a major issue with this particular strain, accounting for the majority (34%) of implant associated infections [12]. The process is initiated through three core steps: bacterial adherence to the implant surface, bacterial colonisation and lastly biofilm formation. The biofilm consists of layers of bacteria covered in a self-produced extracellular polymeric matrix, hence phagocytosis cannot occur and antibiotics have no effect [13]. Bacteria are non-specific with their adherence, attaching to both rough and smooth surfaces, and to different types of materials. Bacterial adherence and subsequent biofilm development is detrimental to the performance of implants, and the ensuing infection can also be a cause of significant morbidity and mortality to implant patients. Therefore, implementing strategies to minimise the likelihood of initial bacterial adherence to the implant surfaces is crucial to prevent bacterial infection [14,15].

Findings in recent research [16,17] shows that surface modification by laser treatment can be used to reduce bacterial infection and improve native cell adhesion. Laser surface treatment, particularly when carried out by fibre laser technology, is a novel technique for implant surface modification [18–20], providing a clean, fast and highly repeatable process. One important characteristic of the laser surface treatment implemented in this study is that the rapid solidification process produces a homogenous surface with little thermal penetration, resulting in little to no distortion [21]. Laser treatment has been shown to improve surface hardness and wear corrosion [22–25]. Other sources of pain after primary total hip replacement (THR), as recorded by the UK National Joint Registry (NJR), include adverse soft tissue reaction to particulates and to infection, both of which can be improved upon by combining better material selection in the form of a novel beta titanium alloy with surface modification using fibre laser treatment. In terms of the economic benefit, TNZT is potentially superior to currently used materials, as it, coupled with laser surface treatment, can simultaneously decrease the most common indications for hip revision surgery listed in the NJR reports [1,26–29]. TNZT use for hip stems may also have a longer life span beyond the current three quarters of hip replacements that last 15–20 years [30] before a revision surgery is required. A systematic review conducted suggested that just over half of hip replacements last 25 years [30]; as ever, there is room for improvement.

If a quantitative relationship between the surface features of an implant and the cell responses were to be established, implants could be designed with specific surfaces which would aid in the host's natural healing processes [31]. To date, very little work has been conducted using the aforementioned composition of beta titanium alloy, and even less specifically focusing on the effect of surface features in relation to MSC response.

The study objectives were to investigate surface feature effects, namely, roughness, topography, composition and chemistry of untreated and fibre laser treated beta titanium alloy Ti-35Nb-7Zr-6Ta on human MSC response by assessing attachment, proliferation and differentiation at various time points, as well as *S. aureus* bacterial attachment, to determine if biocompatibility and antibacterial properties can be improved upon simultaneously using fibre laser treatment.

2. Experimental Section

2.1. Materials

Ti-35Nb-7Zr-6Ta plates were sourced (American Element, Los Angeles, CA, USA) with dimensions of 250 mm × 250 mm with 3 mm thickness, and were wire cut using electrical discharge machining (EDM) (Kaga, Ishikawa, Japan) into 30 mm × 40 mm plates. The material was polished prior to laser treatment using a progression of silicon carbide (SiC) papers with a finish of 1000 grit. Standard metallographic procedures were followed to remove the pre-existing oxide layer and any surface defects present after the manufacturing process. Samples were ultrasonically cleaned in acetone for 10 min, rinsed with deionised water and air-dried prior to laser treatment and material characterisation. A sample size of n = 3 was used for all material characterisation and in vitro cell culture experiments except bacterial attachment, for which n = 4 samples were used.

2.2. Laser Treatment

Laser surface treatment was performed using an automated continuous wave (CW) 200 W fibre laser system (MLS-4030). The laser system was integrated by Micro Lasersystems BV (Driel, Gelderland, the Netherlands) and the fibre laser was manufactured by SPI Lasers UK Ltd (Southhampton, Hampshire, UK). The laser wavelength was 1064 nm. The samples were prepared using the following parameters: laser power 30 W, stand-off distance 1.5 mm, argon gas with 30 L/min flow rate and two different scanning speeds, 100 and 200 mm/s. Laser sample groups are denoted as LT100 and LT200 hereafter. The laser-treated area of the surface was 6 mm^2 in square shape. The laser energy at the two speeds was 1.8 and 0.9 J respectively (see Supplementary Materials for calculations). The control base metal samples (1000 grit finish) are denoted as BM. The sample plate was used for bacterial attachment, otherwise samples were wire cut using EDM into 6 mm diameter discs. Prior to biological culture, samples were ultrasonically cleaned in acetone twice for 1 h, then in deionised water for 30 min and air-dried in the fume hood before a final sterilisation step in an autoclave (Prestige Medical, Blackburn, Lancashire, UK) at 121 °C and 1.5 bar pressure for 20 min to destroy any microorganisms present on the surface.

2.3. Surface Roughness, Topography and Composition

The surface roughness and 3D profile of the untreated and laser treated samples were captured using white light interferometry (WLI) (Talysurf CCI 6000, Leicester, Leicestershire, UK). Roughness was assessed using four parameters: arithmetic mean (Ra), maximum profile height (Rz), surface skewness (Rsk) and surface kurtosis (Rku). Values were extracted from the 1.2 mm^2 scan areas perpendicular to the laser track orientation. Scanning electron microscopy (SEM) was used to image the ripples on the laser treated surfaces (FlexSEM 1000, Hitachi, Maidenhead, Berkshire, UK). SEM images were acquired using a 20 kV beam and backscattered electron compositional (BSE-COMP) mode detection. SEM images for energy dispersive X-ray spectroscopy (EDX) analyses were acquired using a Zeiss Leo 1455VP SEM at 20 kV beam energy with secondary electron detection. EDX data were acquired in the SEM using an Oxford Instruments X-Act detector (Abingdon, UK) with INCA v4.15 acquisition and processing software.

2.4. Phase Identification

Phase and crystallographic structure of the samples were captured by X-ray diffraction (XRD) using a PANanalytical X'Pert Pro MPD (PANalytical, Tollerton, Nottingham, UK) with a CuKα radiation source operated at 40 kV, 40 mA with $\frac{1}{2}°$ fixed slit, 10° anti-scatter slit and 0.02 step size with Ni filter. Samples were analysed in a 2 theta (2θ) range between 10°–90°.

2.5. Surface Chemistry

X-ray photoelectron spectroscopy (XPS) spectra were acquired using a bespoke ultra-high vacuum (UHV) chamber fitted with Specs GmbH Focus 500 monochromated Al Kα X-ray source and Specs GmbH Phoibos 150 mm mean radius hemispherical analyser with 9-channeltron detection. Survey spectra were acquired over the binding energy range between 0 and 1100 eV using a pass energy of 50 eV, and the high resolution scans over the C 1s, Ti 2p, Zr 3d, Nb 3d and O 1s lines were made using a pass energy of 20 eV. Data were quantified using Scofield cross-sections corrected for the energy dependencies of the effective electron attenuation lengths and the analyser transmission. Data processing and curve fitting were carried out using the CasaXPS software v2.3.16 (CasaXPS, Teignmouth, Devon, UK).

2.6. In Vitro Cell Culture

2.6.1. Attachment

Cell culture was performed in a Class II microbiological safety cabinet, and sterile conditions were maintained. Human mesenchymal stem cells (passage 5–9) (Texas A&M Health Science Centre College of Medicine, Institute for Regenerative Medicine, Bryan, TX, USA) were cultured in tissue culture flasks (Thermo Scientific). The medium was comprised of Minimum Essential Medium Alpha with GlutaMAX (Gibco), supplemented with 16.5% foetal bovine serum and 1% glutamine. The cells were maintained in a humidified atmosphere with 5% CO_2 at 37 °C, and were sub-cultured when they reached confluency by washing with phosphate buffered saline (PBS) and disassociated with 0.05% Trypsin-EDTA (Gibco) to provide adequate cell numbers for all studies undertaken, or for further subculture or cryopreservation. Cells were counted using a haemocytometer (Agar Scientific), and seeded in a 96 well plate (Sarstedt) at a density of 5×10^3 cells per well and maintained in the same culture conditions as previously mentioned. Early cell attachment was assessed at the following times: 6, 24 and 48 h using direct immunofluorescent staining, the procedure for which is as follows. Cells were quickly washed with cold PBS, fixed with 4% paraformaldehyde (PFA) for 15 min and permeabilised with 0.1% Triton X-100 in PBS for 5 min. Cells were then blocked with 5% donkey serum in PBS for 30 min. The cells were stained with α smooth muscle actin (SMA)-Cy3 conjugated mouse antibody (1:200) in 5% donkey serum in PBS for 45 minutes at 37 °C. Cells were then counterstained with 4′,6-diamidino-2′-phenylindole dihydrochloride DAPI (1:1000) in PBS for 5 min.

Each step was performed at room temperature unless otherwise stated. Cells were washed with PBS in between every step except after blocking. The last step involved a final wash with distilled water. Samples were transferred to a new well plate for imaging using the Leica DMi8 inverted fluorescence microscope (Leica, Wetzler, Hesse, Germany). Three images were captured per sample at magnification ×100, giving a total of nine images per group at each time point.

2.6.2. Proliferation

Cell proliferation capacity was assessed using the CyQUANT NF Cell Proliferation Assay Kit (Thermo Scientific) at 3, 7 and 14 days. The dye solution was prepared by adding 1× dye binding solution to 1× Hank's balanced salt solution (HBSS). The cell culture medium was aspirated and 200 µl of the dye solution was added to each well and incubated at 37 °C for 30 min. The fluorescence intensity was measured using a microplate reader (Varioskan LUX) at 485 nm excitation and 530 nm emission.

2.6.3. Differentiation

Cells were stained using indirect immunofluorescent assay staining at 7, 14 and 21 days. All steps were the same as previously mentioned, see Section 2.6.1, except after blocking, where cells were instead incubated with anti-osteocalcin (10 µg/ml) (R&D Systems) in 5% donkey serum in PBS for 1 h. Cells were then stained with Alexa Fluor 488 donkey anti-mouse (1:500) (Thermo Scientific) in PBS for 45 min. This was followed by DAPI staining, PBS wash and distilled water wash. Samples were transferred to a new well plate for imaging using the Leica DMi8 inverted fluorescence microscope (Leica, Wetzler, Hesse, Germany). Three images were captured per sample at magnification ×100, giving a total of nine images per group at each time point.

2.7. Bacterial Attachment

The TNZT plate was washed three times with sterile PBS. *S. aureus* (ATC 44023) was cultured in Müller Hinton broth (MHB) for 18 h at 37 °C on a gyrotatory incubator with shaking at 100 rpm. After incubation, sterile MHB was used to adjust the culture to an optical density of 0.3 at 550 nm, and it was diluted (1:50) with fresh sterile MHB. This provided a bacterial inoculum of approximately 1×10^6 colony forming units (CFU)/mL. 1 mL of culture was applied to the plate carefully suspended over a petri dish base at an inoculum not exceeding 2.4×10^6 CFU/mL, as verified by viable count. The plate was incubated for 24 h at 37 °C on a gyrotatory incubator with shaking at 100 rpm. Four samples of each group, for both untreated and laser treated, were tested to ensure the consistency of the results. After 24 h of incubation, the plate was washed three times with sterile PBS to remove any non-adherent bacteria. The adherent bacteria were stained with fluorescent Live/Dead®BacLight™ solution (Molecular Probes) for 30 min at 37 °C in the dark. The fluorescent viability kit contains two components: SYTO 9 dye and propidium iodide. The SYTO 9 labels all bacteria, whereas propidium iodide enters only bacteria with damaged membranes. Green fluorescence indicates viable bacteria with intact cell membranes, while red fluorescence indicates dead bacteria with damaged membranes. The stained bacteria were observed using a fluorescence microscope (GXM-L3201 LED, GX Optical, Stansfield, Suffolk, UK). Twenty four random fields of view (FOV) were captured per group. The surface areas covered by the adherent bacteria were calculated using ImageJ software. The areas corresponding to the viable bacteria (coloured green) and the dead bacteria (coloured red) were individually calculated. The total biofilm area was the sum of the green and red areas, and the live/dead cell ratio was the ratio between the green and red areas. The results are expressed as the means of the twenty four measurements taken per group.

2.8. Statistical Analysis

Data were expressed as mean ± standard error (SE). The significance of the observed difference between untreated and laser treated samples was tested by one-way and two-way ANOVA using Prism software (GraphPad Prism Software Version 7, San Diego, CA, USA). Statistically significant differences were calculated, with a p-value of ≤ 0.05 considered significant; (*) $p < 0.05$; (**) $p < 0.01$; (***) $p < 0.001$.

3. Results

3.1. Surface Roughness by WLI

The 3D and 2D profiles of the untreated and laser treated surfaces can be seen in Figure 1A,B. The laser treated surfaces appear to have very smooth tracks (as indicated by each individual colour scale), showing a higher degree of surface polishing effect with increased laser scanning speed, as shown by the shift in the colour scale of the 2D profiles. It is also evident that the base metal areas in between the laser tracks have significant variation in surface roughness, as is expected of the metallographic preparation process (i.e., polished by SiC paper). The roughest regions are prominent at the very edge of the laser tracks for each surface (as indicated by the arrows in Figure 1A,B).

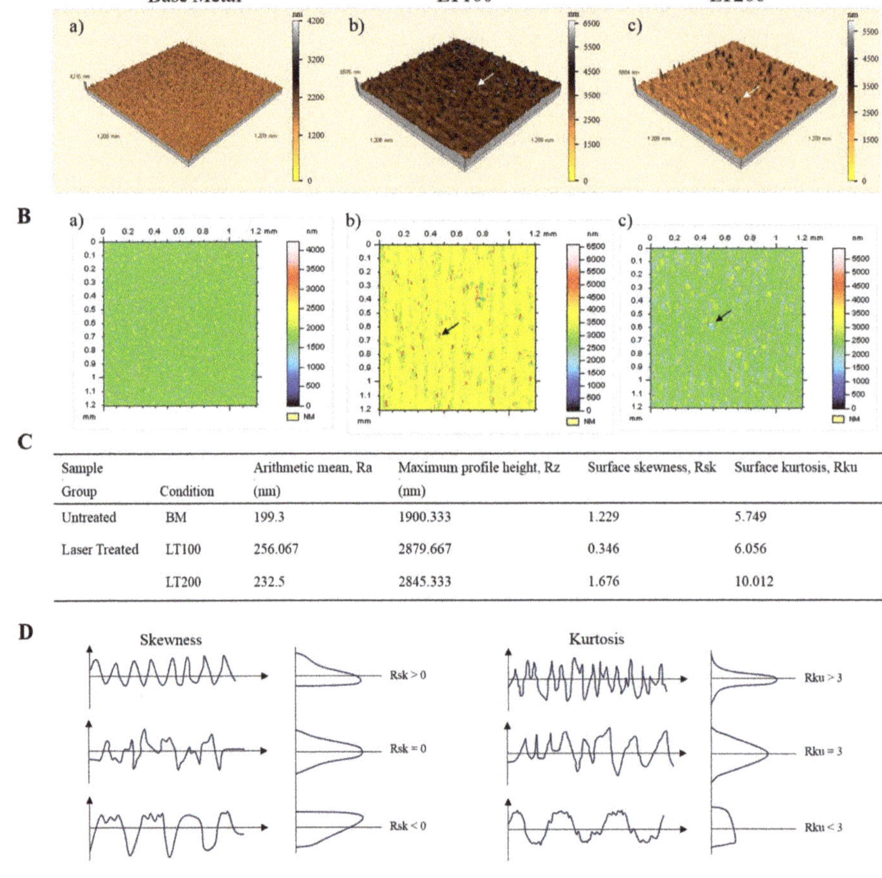

Sample Group	Condition	Arithmetic mean, Ra (nm)	Maximum profile height, Rz (nm)	Surface skewness, Rsk	Surface kurtosis, Rku
Untreated	BM	199.3	1900.333	1.229	5.749
Laser Treated	LT100	256.067	2879.667	0.346	6.056
	LT200	232.5	2845.333	1.676	10.012

Figure 1. Laser surface treatment has a polishing effect, which increases with laser scanning speed while creating a rough and spiky laser track edge. White light interferometry (**A**) 3D and (**B**) 2D profiles for the untreated and laser treated samples. Scan area of 1.208 mm^2 and 1.2 mm^2 for 3D and 2D images respectively; colour coding of each image seen by individual colour scale in figure. Arrows show the spiky areas at the laser track edges. (**C**) Summary of surface roughness parameters. (**D**) Diagrams illustrating visual difference between surface skewness and kurtosis profiles.

The surface roughness values extracted from the 2D profiles can be seen in Figure 1C. The Ra values ranged from a low of 199.3 nm on BM to a highest value of 256.1 nm on the LT100 surface, while LT200 had a Ra of 232.5 nm. The Rz values followed a similar trend to the Ra, with BM having the lowest maximum profile height (1900.3 nm) while LT100 and LT200 had similar values, 2879.7 and 2845.3 nm, respectively. Overall, the laser surface treatment increased the Ra and Rz in comparison to the untreated surface, due to the rough edges created on the edges of the laser created tracks.

The untreated and laser treated samples were also quantified using the surface skewness and kurtosis values. The skewness defines whether a surface consists of spikes (Rsk > 0) or valleys (Rsk < 0), while the kurtosis defines whether a surface has peaks (Rku > 3) or is flat (Rku < 3); see Figure 1D for diagram illustrating the visual differences between varying Rsk and Rku values. There was a slight increasing trend in Rsk and a more notable difference in Rku as the laser scan speed increased, as seen in Figure 1C. The LT100 surface had the lowest Rsk (0.346), while the BM had a value of 1.229. The LT200 had the highest Rsk with 1.676, showing that the highest scanning speed creates the spikiest surface. All Rku values were >3, with BM being the lowest at 5.749 and increasing with scanning speed. There was a sharp increase in Rku at the highest scanning speed, reinforcing the laser surface polishing effect. All the surfaces have a Rsk > 0 and Rku > 3. The LT200 group had the highest skewness and kurtosis, suggesting it had more peaks and spikes present on the edges of the laser created tracks, but the smoothest laser treated area.

3.2. Surface Topography and Composition by SEM-EDX

The SEM images of the untreated and laser treated surfaces can be seen in Figure 2A(a–d). The typical surface morphology after mechanical grinding can be seen above and below each laser track in Figure 2A(a,b), with the surface exhibiting random scratches, pits and grooves. The tracks created using laser treatment had a distinctive ripple effect, which was more prominent at the lower scan speeds, with small ripples present along the entire track and the periodic appearance of distinctive larger ripples. The magnified SEM images show that the ripples were much smaller and uniform on the LT200 surface than on LT100, which had more distinct arches, as shown in Figure 2A(c,d). The tracks created by laser surface treatment became smoother as the scanning speed increased, verifying the laser surface polishing effect observed in the WLI images.

The ripple effect was quantified by calculating the peak-to-peak distance between ripples, using Image J, as seen in Figure 2B,C. The distance became smaller with increasing laser scanning speeds. The peak-to-peak distance between ripples halved when comparing LT100 and LT200, dropping from 0.63 to 0.315 μm. The surfaces were significantly different from one another ($p < 0.001$).

The SEM-EDX analyses were performed to determine if there were any notable differences in elemental distribution post laser treatment, and can be seen in Figure 2D. The base metal polished areas were quite distinctive from the smooth laser treated tracks; the boundary between these areas is defined by a dashed yellow line. The carbon and oxygen were quite densely concentrated on the base metal polished area of the samples, while homogeneously distributed in the laser tracks on each surface, irrespective of scanning speed. It is apparent that for each element, except Ta, the laser treatment created a homogenous elemental distribution within the tracks with no measureable spatial variation. Enriched particles of Ti, Nb and Zr can be seen in Figure 2D(i–vi) for each surface, as indicted by the black arrows, which can be attributed to the base metal polished surface features. There was a more uniform and homogeneous oxide film present post laser treatment.

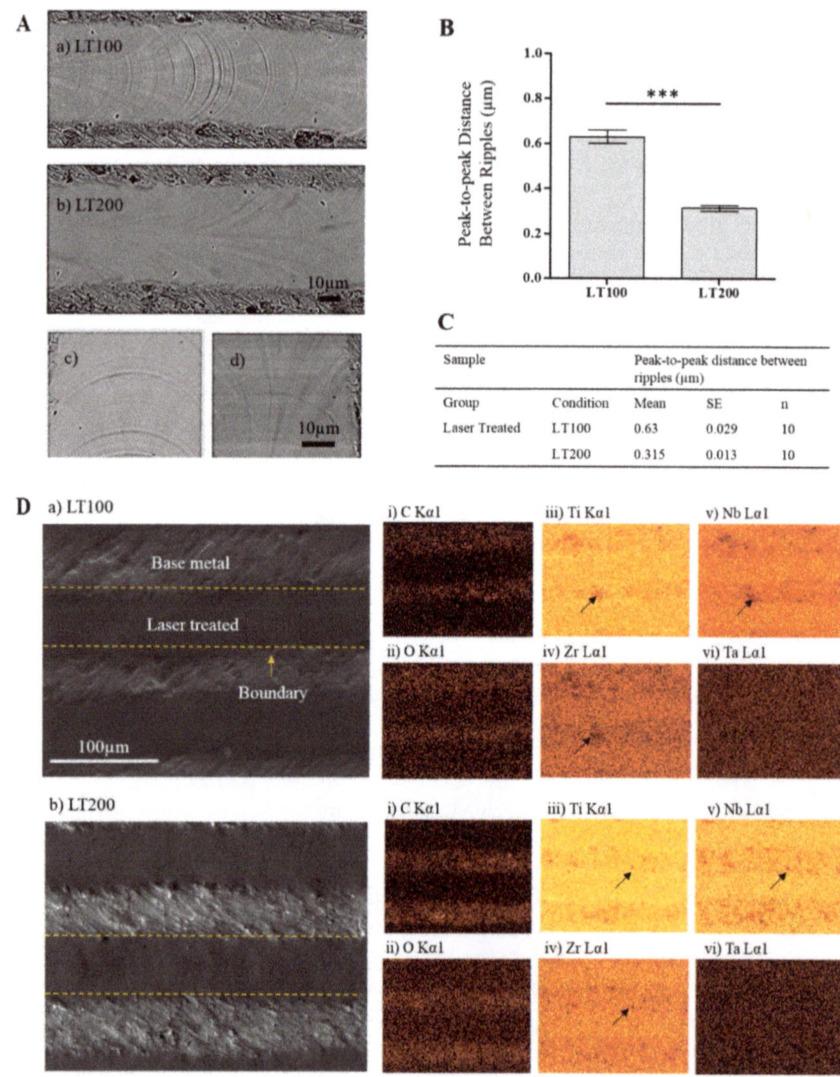

Figure 2. Distinctive microscale ripples and homogenous elemental distribution achieved through laser surface treatment. (**A**) SEM images of the multi-scale ripples created within the laser tracks. Scale bar = 10 μm. (**B**) Peak-to-peak distance between ripples, calculated using ImageJ. Error bars are the SE of n = 10. (**C**) Tabulated peak-to-peak distance between ripples. (**D**) SEM-EDX images showing the laser treated surfaces, with a boundary defining the base metal polished and laser treated areas. Scale bar = 100 μm. EDX elemental maps show the spatial distribution of carbon, oxygen, titanium, niobium, zirconium and tantalum. Black arrows indicate particle enrichment in the base metal polished zones for all elements except tantalum.

3.3. Phase Identification by XRD

The phase and structure of the samples were identified using XRD, as seen in Figure 3. There was a notable preferential crystallographic phase shift post laser treatment. The blue dot indicates the presence of a specific peak in all samples. The untreated base metal surface had a prominent peak at ~56° β (200), with additional smaller peaks present (~38.5° β (110) and ~69° β (211)). The crystallographic plane shifted to ~38.5° β (110) after laser treatment for LT100. All samples had a peak present at ~38.5° β (110), ~56° β (200) and ~69° β (211), although LT100 had an additional weak peak present at ~83° (β 220). All peaks were associated with beta phases, no alpha phases were detected. This is due to the presence of beta stabilising elements (niobium and zirconium) in the material, which suppresses the formation of alpha phase.

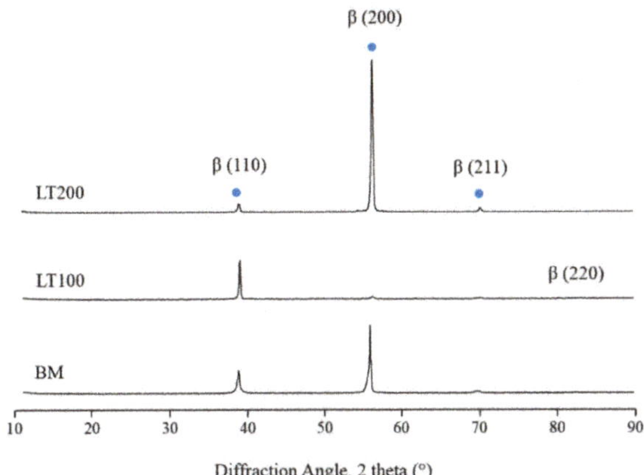

Figure 3. Shifting crystallographic plane with laser surface treatment. XRD profiles of the untreated and laser treated samples. The scanning diffraction angles are between 10° and 90°. Y-axis shows intensity (a.u.).

3.4. Surface Chemistry by XPS

The XPS spectra and narrow scans for the untreated and laser treated groups can be seen in Figure 4. A summary of the % concentrations by XPS on untreated and laser treated surfaces can be found in Table 1, and a summary of the species detected are noted in Table 2. All surfaces showed the expected Ti, Nb, Zr, C and O, with the additional presence of N, Zn and Cu. The presence of organic nitrogen is probably due to sample exposure to air [32]. The Cu and Zn on the BM surface were attributed to surface contamination during manufacture of the alloy, and showed levels which reduced as a consequence of laser treatment. Each element and assignment, as seen in Table 2, was found in all of the laser treated samples.

Table 1. Summary of % concentrations by XPS on untreated and laser treated surfaces.

	XPS (% Concentration)		
Name	**BM**	**LT100**	**LT200**
C 1s	54.1	56.4	52.2
N 1s	7.7	4.4	3.2
O 1s	27.4	30.8	32.9
Si 2p	1.8	0.5	0.6
Ti 2p	3.4	6.1	8.6
Cu 2p3/2	2.7	0.2	0.8
Zn 2p3/2	0.8	0.6	0.4
Zr 3d	0.4	0.5	0.7

The untreated sample had no Ti $2p_{3/2}$ in the form of Ti metal present in the surface layer, and a very weak presence of Ti^{3+} in Ti_2O_3/TiN at 456.2 eV. The majority of Ti at the outermost untreated surface was found in the Ti^{4+} state in TiO_2 at 458.4 eV. The case was similar for the two other metal elements, niobium and zirconium. Nb was only found in the Nb^{5+} state in Nb_2O_5 at 207 eV, while the majority of Zr was found in the Zr^{4+} assignment in ZrO_2 at 182.3 eV. There was very little to no nitride present on the untreated sample, whereas there was some evidence from the laser treated samples to suggest that the laser treatment creates a nitride layer at a BE range of 395.8–397 eV.

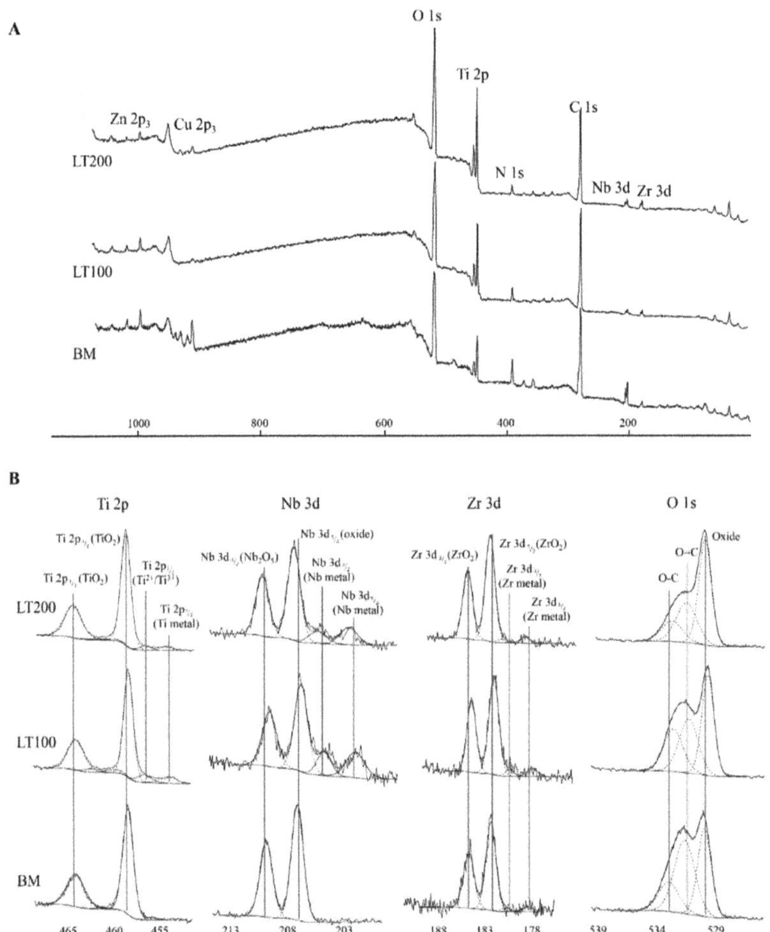

Figure 4. XPS spectra of untreated and laser treated samples. (**A**) XPS spectra: Zn 2p3, Cu2p3, O 1s, Ti 2p, N 1s, C 1s, Nb 3d and Zr 3d. (**B**) XPS narrow scan spectra: Ti 2p, Nb 3d, Zr 3d and O 1s for untreated and laser treated surfaces. X-axes show binding energy (eV).

The oxygen spectra for the untreated and laser treated samples typically showed a relatively sharp lower binding energy metal-bonded component, and a broader but relatively featureless higher binding energy component typical of organic oxygen, e.g., bonded within the hydrocarbon contamination layer. The metal-bonded component (i.e., Ti–O, Nb–O and Zr–O) was typically found at 529.8 eV. The carbon–oxygen region was typically fitted with two components at 530.9–531.5 eV and 532.3–532.9 eV, representative of O=C and O–C bonding respectively, typically associated with residual organic contamination.

Table 2. Summary of the chemical species detected by XPS.

Element	Line	Assignment	BE Range (eV)	Present in Sample		
				BM	LT100	LT200
Ti	$2p_{3/2}$	Ti metal	454.3	No	Yes	Yes
		Ti^{3+} in Ti_2O_3/TiN	456.2	V. weak	Yes	Yes
		Ti^{4+} in TiO_2	458.4	Yes	Yes	Yes
O	1s	Ti–O, Nb–O, Zr–O	529.8	Yes	Yes	Yes
		O=C	530.9–531.5	Yes	Yes	Yes
		O–C	532.3–532.9	Yes	Yes	Yes
C	1s	C–C	285	Yes	Yes	Yes
		C–O	286–286.4	Yes	Yes	Yes
		C=O	287.2–288.1	Yes	Yes	Yes
		COO-	288.6–289.3	Yes	Yes	Yes
Nb	$3d_{5/2}$	Nb metal	202–202.5	No	Yes	Yes
		Nb^{5+} in Nb_2O_5	207	Yes	Yes	Yes
Zr	$3d_{5/2}$	Zr metal	178.4	Weak	Yes	Yes
		Zr^{4+} in ZrO_2	182.3	Yes	Yes	Yes
N	1s	Nitride	395.8–397.0	V. weak/No	Yes	Yes
		Organic	399.7–400.3	Yes	Yes	Yes

3.5. Cell Responses

3.5.1. Attachment

It is evident at the early attachment time points that the cells behaved distinctively differently on the laser treated surfaces in comparison with the untreated surfaces between the 24 and 48 h time points, as seen in Figure 5. Alpha SMA was specifically used to clearly visualise cell morphology. The MSCs were visually similar in shape at 6 h on all surfaces, displaying the typical polygonal structure with uniform spreading regardless of underlying surface topography. At 24 and 48 h, the cells on the laser treated surfaces began to show evidence of interacting with the surface. At 24 h, the LT100 cells remained fairly rounded and polygonal in shape, while the LT200 surfaces encouraged cell stretching, seen across and along the laser created tracks. At 48 h, the cell shapes had vastly changed again, with LT100 causing the cells to become slightly smaller in size in comparison with their untreated BM counterparts. LT200 appears to have the most influence on cell shape, as it is clear that at 48 h the cells displayed a spindle shaped appearance. Meanwhile, the BM surface encourages cells to stretch. This is probably an effect of the scratch marks remaining after the SiC paper polishing process.

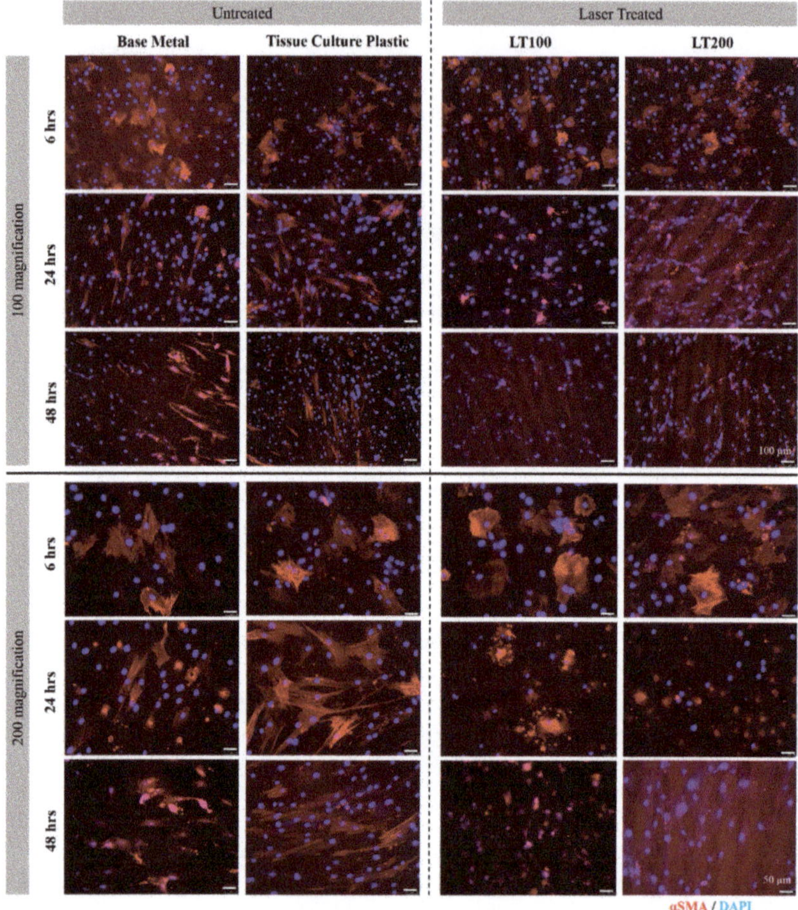

Figure 5. Laser surface treatment at high scanning speeds can be used to encourage cell spreading. Merged fluorescence images of mesenchymal stem cells (MSCs) stained at 6, 24 and 48 h at 100× (top) and 200× (bottom) magnification on untreated and laser treated surfaces. Red = αSMA actin fibres; blue = DAPI nuclei stain. Scale bar = 100 μm (top) and 50 μm (bottom). Images are representative of cell coverage on the entire surface.

3.5.2. Proliferation

The fluorescence intensity of MSCs on the untreated and laser treated surfaces at Day 3, 7 and 14 can be seen in Figure 6. A higher intensity is associated with more cells being present on the surface. There were no significant differences between surfaces within time points until Day 7 and 14. At Day 7, there was a significant difference between BM and LT100 ($p < 0.05$). At Day 14, the untreated BM had the highest fluorescence intensity, followed by LT200 then LT100. BM was significantly different from the two laser treated groups, LT100 ($p < 0.001$) and LT200 ($p < 0.05$).

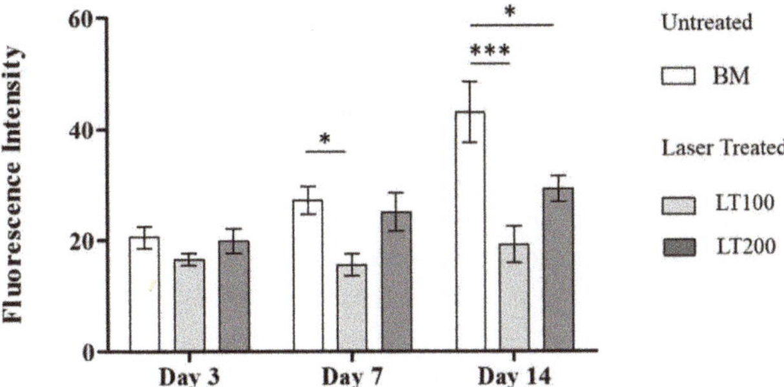

Figure 6. Fluorescence intensity of MSCs at Day 3, 7 and 14 proliferation time points on untreated and laser treated surfaces. The error bars indicate the SE of n = 3.

3.5.3. Differentiation

MSC morphology was qualitatively assessed using fluorescence staining of osteocalcin at Day 7, 14 and 21. There was a very distinct difference in MSC morphology on the untreated and laser treated surfaces, as seen in Figure 7. Those on BM were spindle shaped, while TCP had good coverage, as expected. More distinct cell shapes can be seen on LT200 than on LT100. The distinctly different cell morphology can be seen at the later differentiation time points. The number of cells increased on all surfaces between Day 7 and Day 14, with TCP having nearly full coverage and monolayer. The cells on BM had the same spindle shape, while the cells on both laser treated surfaces had begun to form clusters, the larger seen on LT200, with small rounded cells present alongside the clusters.

At Day 21, the TCP had formed a monolayer of MSCs, while the cell clusters could be seen on the BM surface. There were multiple cell cluster formations on the laser treated surfaces, which could be indicative of bone-like nodule formation, suggesting that the parameters chosen for the laser surface treatment encourage bone to grow more quickly than on the BM surface, suggesting that laser surface treatment is a potential modification technique to encourage faster bone growth.

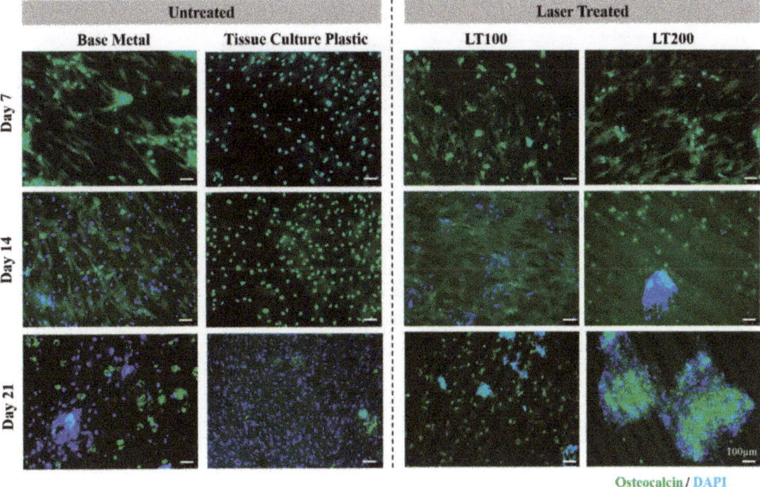

Figure 7. Laser surface treatment encourages formation of bone-like nodules, indicating potential faster osseointegration. Merged fluorescence images of MSCs stained at Day 7, 14 and 21 at 100× magnification on untreated and laser treated surfaces. Green = osteocalcin protein; blue = DAPI nuclei stain. Scale bar = 100 μm. Images are representative of cell coverage on the entire surface.

3.6. Bacterial Attachment

The bacterial attachment results on the untreated and laser treated surfaces can be seen in Figure 8. Attachment was quantitatively analysed using *S. aureus* coverage and live-to-dead ratio to determine which surface(s) elicited a bactericidal response. The fluorescence images show that live cells were green stained with SYTO 9, while dead cells were red stained with propidium iodide. Green fluorescence indicates viable bacteria with intact cell membranes, while red fluorescence indicates dead bacteria with damaged membranes. As seen in Figure 8A, there was a visibly higher number of green stained cells present on the untreated surface. After only 24 h, the lower number of bacteria on the laser treated surface in comparison with the untreated suggests that laser treatment created an inhospitable environment for the bacteria, causing them to become non-viable. The bacterial attachment results were quantified using bacteria coverage and the ratio of live-to-dead cells. The bacterial coverage on each laser treated surface was significantly different ($p < 0.001$) from the untreated BM surface, see Figure 8B. There was a four-fold increase in the bacterial coverage between the laser treated and untreated group. Likewise, the live/dead ratio of bacteria present on the laser treated surfaces was also drastically reduced, see Figure 8C.

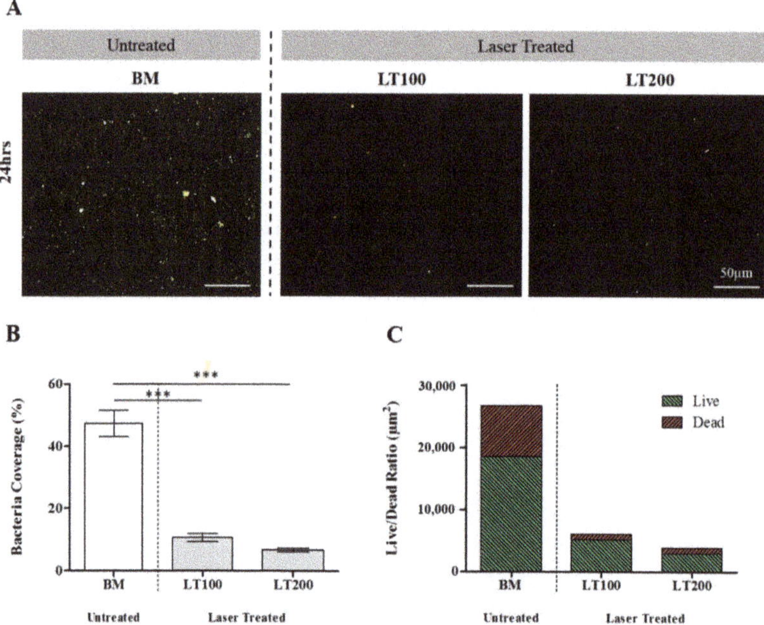

Figure 8. Bacterial adherence significantly reduced with laser surface treatment. (**A**) Merged fluorescence images of bacteria at 24 h using live/dead staining. Scale bar = 50 μm. Images are representative of bacteria coverage on the entire surface. (**B**) Total bacteria coverage measured from fluorescence microscopy. Error bars are the SE of n = 24. The total bacteria coverage refers to the total sum of red and green areas present after cell staining. (**C**) Live-to-dead ratio of bacteria cells measured from fluorescence microscopy on the surfaces post staining, based on average red and green areas per group. Data calculated in both graphs using twenty four images per surface at each time point using ImageJ software.

4. Discussion

Bimodal texturing is an important consideration for improvement in surface feature design. Micro and nanotopography must be used in tandem to create a surface that is sufficient for long-term stability [19,33]. Micro scale features such as grooves, ridges and pits can increase surface area and provide more opportunities for attachment, and these features can cause cells to align and organise within them. Features at the nano scale directly affect protein interactions, filaments and tubules, which control cell signalling and regulates cell adhesion, proliferation and differentiation [34].

The roughest regions were prominent at the very edge of the laser tracks for each surface (as indicated by the arrows in Figure 1A,B). This can be attributed to the consequence of melt pool dynamics at the liquid/solid boundary (or melt pool/heat affected zone (HAF) boundary) during the laser re-melting process [35]. It is known that the micro-ripple surface (as seen in Figure 2A) results from oscillation of the liquid metal due to the Marangoni convection and hydrodynamic processes driven by thermocapillary motion acting on the melt pools [36,37]. The black marks seen in Figure 2A(a–d) could be due to contamination from the material handling process, although further in-depth analysis is required.

There was a more uniform and homogeneous oxide film present post laser treatment (Figure 2D). The elements on the LT100 and LT200 surfaces were uniformly distributed with no measureable spatial variation over the surface area imaged, suggesting the elongated and thin brush-like marks (extended from the interior areas of laser tracks to the boundary, as seen in Figure 2A) were laser-induced surface features as a consequence of the convection field in the complex melt pool dynamics and solidification processes during laser treatment [38]. Tantalum is the only element that did not show any evidence of particle enrichment, perhaps due to it being the least abundant element present in the quaternary alloy.

There was a notable preferential crystallographic phase shift post laser treatment (Figure 3). This could be due to the preferential orientation of a specific phase (i.e., peak angle of ~38.5°) caused by higher laser energy input at lower scanning speed. However, further in-depth analysis is still required to investigate how the difference of laser energy input between the two scanning speeds (i.e., lower at 100 mm/s and higher at 200 mm/s) makes the change. Sharp dominant phase peaks present in XRD analysis indicated the treated material had a high degree of crystallinity.

The significant reduction of bacterial attachment and/or biofilm formation on the laser treated surfaces can be attributed to the following: (i) Firstly, the SEM-EDX results indicated "homogenisation" or "finer dispersion" of metal compositions in the laser-melted surfaces (i.e., absence of metal-enriched particles, as indicated in Figure 2D). In addition, the laser treated surfaces look smoother in the SEM images (Figure 2A), and the original fine-scale texture and roughness of the untreated areas was not present after laser treatment. Both of these can be linked to the formation of a more uniform and homogeneous oxide film after laser treatment. (ii) Secondly, the XPS data and also the SEM-EDX results indicate that overall the oxide film was thinner after laser treatment, but it was likely to be more uniform in thickness, as described in point (i) above. Therefore, possibly, there was a better coverage of a more well-defined though thinner oxide layer over the treated areas. (iii) Thirdly, removal of residual organic contaminants, as indicated by the reduced proportion of both oxygen-bonded and carbon-bonded species in the oxide film, which could be acting as potential sources to attract bacteria to attach on surfaces via non-covalent interactions [39]. The phenomenon of laser treatment helps reduce overall levels of organic surface contaminants, as reported elsewhere [16,36]. This can be due to rapid vaporisation of more volatile species under the sudden input of energy from the laser, or, possibly, more adherent contaminants are buried as the locally-melted metal re-solidifies after laser treatment. (iv) Finally, the presence of titanium nitride (TiN) in the oxide film after laser treatment could contribute to the antibacterial activities. It has been reported that a surface with TiN can deactivate biofilm formation [14,40]. Likewise, zirconium nitride (ZrN) is known to be an antibacterial material [41]. However, the possibility of an antibacterial effect attributed to ZrN can be eliminated in this study, because there is no evidence for the existence of ZrN in the oxide film, as shown in the XPS narrow scan profile in Figure 4B (i.e., the binding energy for ZrN would be expected to be around 180.9 eV [42]). In contrast, the evidence for TiN present in the oxide film after laser treatment is clear (i.e., the curve fitted at the binding energy around 456.2 eV in Figure 4B).

It is important to note that, although the evidence for the appearance of metallic species in the oxide film, namely Ti and Nb metals, after laser treatment is also very clear, the results in this study indicated that they are not necessarily encouraging the bacterial attachment and/or biofilm formation. The presence of the oxide layer improves the corrosion resistance of the material's passive surface [43]. The organic contamination found in the form of O–C and O=C bonds could be due to the material handling process or a carbon-containing cleaning agent; further in-depth analysis is required.

The ultrastructure of the bone–titanium interface demonstrates simultaneous direct bone contact, osteogenesis and bone resorption [44]. Upon implantation, a material surface initially interacts with water, followed by protein adsorption then cell interaction, among which is included MSCs. The surface macro scale is responsible for the interlock between bone and implant [45]. The micro scale can influence cell orientation [46] and potentially proliferation capacity and differentiation ability. The nano scale can influence cell-to-cell signalling [34], and can override biochemical cues [47]. Independent of the surface chemistry, the surface scale, namely micro and nano topography, has a significant effect on cell behaviour [48]. The cell cytoskeleton organisation is strongly affected by the orientation of the surface structures (i.e., physical roughness and topography), which stimulates cell contact guidance [49]. Contact guidance refers to the phenomenon where cells will adjust their orientation and align along nano-micro-groove-like patterns to grow. It was first observed by Harrison [50] in 1912, and the terminology first described by Weiss and Taylor [46] in 1945. Curtis and Wilkinson reviewed the materials (one being titanium surface oxides) and topographical structures which can effect cell behaviour, such as grooves, ridges, spikes and pits [51]. Research has evolved since, and now emphasis is placed on how cell geometric cues can direct cell differentiation, and in the process manipulate cells into square, rectangular and pentagon shapes [52].

The cell coverage was similar across the early attachment time points (Figure 5) which correlates with the proliferation data at Day 3 (Figure 6). Although the coverage was similar, the morphology was distinctive on the laser treated surfaces. At Day 14, proliferation results show that BM surface had the highest fluorescence intensity (significantly different from both laser treated surfaces), although at Day 14 the differentiation results show that BM had spindle shaped cells and approximately 60% coverage, while the laser treated groups first began to show morphological evidence of cluster formation, perhaps indicative of bone-like nodules, although additional in-depth analysis is required to further characterise the cell behaviour. It is clear that cell shape is a relevant parameter in the biomaterial design process, as a fundamental physiological feature of functional tissue [53]. Faster bone formation by laser surface treatment could be explained by the hypothesis that osteoblast precursors migrating into the pores of a rough surface reach confluence earlier within the enclosed space, cease proliferation and then differentiate [54]. Improvement of overall in vitro performance links to the surface roughness and topographical features, with these being the main indicators of osseointegration success [19,55,56]. The introduction of laser technology for titanium surface modification is feasible, and evidently beneficial for accelerating bone formation [57].

The antibacterial effect arising from the changes in surface chemistry of TNZT after laser treatment could apply to the attachment of MSCs, i.e., negatively impacting the MSCs attachment on laser treated surfaces. However, positive results of attachment and coverage of MSCs on the laser treated surfaces, at least comparable with that of the untreated (polished) surfaces, can still be observed in this study. It can be attributable to the size difference between bacteria and MSCs, namely 0.5–1 μm and 20–30 μm, respectively.

In the authors' recent study [16,36], bacteria were found to be insensitive to the micro-sized surface features, namely micro-ripples in the laser tracks [36], while the response of bacteria was very much dictated by the nano-sized features, i.e., nano-spiky features are effective to inhibit bacterial attachment and to kill the bacteria that attached [16]. Similarly, Puckett et al. found that certain nanometre sized titanium topographies may be useful for reducing bacteria adhesion while promoting bone tissue formation [58]. In contrast, the relatively large-sized MSCs, compared with the bacteria, are more sensitive to surface features in both the micro- and nano-sized range. It has been reported by Chan et al. [59] that laser-induced surface ripples or patterns in micro size can encourage higher cell attachment of MSCs, leading to a higher cell coverage on the surfaces after laser treatment. Surface modification is emerging as a promising strategy for preventing biofilm formation on abiotic surfaces [60]. Implant success relies upon the surface inhibiting bacterial adherence and concomitantly promoting tissue growth.

To summarise, there is a competing process between effects caused by the laser-induced surface chemistry and micro-features, i.e., changes in surface chemistry after laser treatment could make the surface less hospitable to MSCs, whereas the micro-sized ripples (or physical features) can promote more cell attachment and coverage. However, at this stage it is still inconclusive as to how the laser-induced chemical or physical effects individually act on the response of MSCs, or which one is more dominant. A single surface feature effect cannot be studied in isolation from the others.

5. Conclusions

The beta titanium alloy Ti-35Nb-7Zr-6Ta coupled with fibre laser surface treatment in a high speed regime (ranging 100 and 200 mm/s) is a promising choice for load bearing implant applications. The surface roughness, topography and composition can be tailored by fibre laser treatment to improve in vitro mesenchymal stem cell attachment, proliferation and differentiation, as well as reducing bacterial attachment.

The major findings of this research are summarised below:

(1) Fibre laser treatment can be used to polish the TNZT surfaces in the high speed regime, with the scanning speed of 200 mm/s (or 12 m/min) being the most effective in this study;

(2) The laser treated samples exhibited surface homogenisation (or homogenous elemental distribution), and only showed beta phases after fibre laser treatment, namely β (110) and (200);

(3) Fibre laser treatment created a rougher (Ra value of BM was 199 nm, LT100 was 256 nm and LT200 was 232 nm) and spiky surface (Rsk > 0 and Rku > 3) with a homogenous elemental distribution and presence of TiN in the outmost oxide layer, which encouraged bone-like nodule formation and a bactericidal effect.

To summarise, the cell (attachment, proliferation and differentiation of MSCs) and bacterial culture (live/dead ratio of *S. aureus*) results indicate that LT200 is the optimal condition to treat the TNZT surface, giving the most desirable MSC responses and a significant reduction in bacterial adhesion.

Supplementary Materials: The following are available online at http://www.mdpi.com/2079-6412/9/3/186/s1.

Author Contributions: Conceptualization, C.L.D. and C.-W.C.; methodology, C.L.D., G.C.S. and L.C.; formal analysis, C.L.D., R.M., G.C.S., A.M. and C.-W.C.; investigation, C.L.D.; resources, R.M., G.C.S., S.K., L.C., S.M., A.M. and C.-W.C.; data curation, C.L.D. and C.-W.C.; writing—original draft preparation, C.L.D., G.C.S. and C.-W.C.; writing—review and editing, C.L.D., R.M.F., G.C.S., L.C., A.M. and C.-W.C.; visualization, C.L.D. and C.-W.C.; supervision, S.M., A.M. and C.-W.C.; project administration, C.-W.C.; funding acquisition, L.C., A.M. and C.-W.C.

Funding: The work described in this paper was supported by research grants from the Queen's University Belfast, awarded to C-WC and LC (D8201MAS, D8304PMY). The cell work was funded by Biotechnology and Biological Sciences Research Council (BBSRC) and British Heart Foundation (BHF) grants.

Acknowledgments: Anna Krasnodembskaya's group at the Centre for Experimental Medicine (Queen's University Belfast) for the provision of MSCs.

Conflicts of Interest: The authors declare no conflict of interest.

References

1. National Joint Registry. *National Joint Registry for England, Wales and Northern Ireland 14th Annual Report 2017*; Pad Creative Ltd.: London, UK, 2017.

2. Miyamoto, T.; Suda, T. Differentiation and function of osteoclasts. *Keio J. Med.* **2003**, *52*, 1–7. [CrossRef] [PubMed]

3. Abu-Amer, Y.; Darwech, I.; Clohisy, J.C. Aseptic loosening of total joint replacements: Mechanisms underlying osteolysis and potential therapies. *Arthritis Res. Ther.* **2007**, *9*, 56–62. [CrossRef] [PubMed]

4. Liu, X.; Chu, P.K.; Ding, C. Surface modification of titanium, titanium alloys, and related materials for biomedical applications. *Mater. Sci. Eng. R Rep.* **2004**, *47*, 49–121. [CrossRef]

5. Niinomi, M.; Nakai, M.; Hieda, J. Development of new metallic alloys for biomedical applications. *Acta Biomater.* **2012**, *8*, 3888–3903. [CrossRef] [PubMed]

6. Kuroda, D.; Niinomi, M.; Morinaga, M.; Kato, Y.; Yashiro, T. Design and mechanical properties of new beta type titanium alloys for implant materials. *Mater. Sci. Eng.* **1998**, *243*, 244–249. [CrossRef]

7. Ku, C.; Pioletti, D.P.; Browne, M.; Gregson, P.J. Effect of different Ti–6Al–4V surface treatments on osteoblasts behaviour. *Biomaterials* **2002**, *23*, 1447–1454. [CrossRef]

8. Akahori, T.; Niinomi, M.; Nakai, M.; Kasuga, T.; Ogawa, M. Characteristics of Biomedical Beta-Type Titanium Alloy Subjected to Coating. *Mater. Trans.* **2008**, *49*, 365–371. [CrossRef]

9. Niinomi, M. Biologically and Mechanically Biocompatible Titanium Alloys. *Mater. Trans.* **2008**, *49*, 2170–2178. [CrossRef]

10. Hutmacher, D.W.; Schantz, J.T.; Lam, C.X.; Tan, K.C.; Lim, T.C. State of the art and future directions of scaffold-based bone engineering from a biomaterials perspective. *Tissue Eng. Regen. Med.* **2007**, *1*, 245–260. [CrossRef]

11. Abdel-Hady Gepreel, M.; Niinomi, M. Biocompatibility of Ti-alloys for long-term implantation. *J. Mech. Behav. Biomed. Mater.* **2013**, *20*, 407–415. [CrossRef]

12. Campoccia, D.; Montanaro, L.; Arciola, C.R. The significance of infection related to orthopedic devices and issues of antibiotic resistance. *Biomaterials* **2006**, *27*, 2331–2339. [CrossRef] [PubMed]

13. Singh, S.; Singh, S.K.; Chowdhury, I.; Singh, R. Understanding the Mechanism of Bacterial Biofilms Resistance to Antimicrobial Agents. *Open Microbiol. J.* **2017**, *11*, 53–62. [CrossRef] [PubMed]

14. Scarano, A.; Piattelli, M.; Vrespa, G.; Caputi, S.; Piattelli, A. Bacterial adhesion on titanium nitride-coated and uncoated implants: An in vivo human study. *Oral Implantol.* **2003**, *29*, 80–85. [CrossRef]

15. Cunha, A.; Elie, A.-M.; Plawinski, L.; Serro, A.P.; do Rego, B.; Maria, A.; Almeida, A.; Urdaci, M.C.; Durrieu, M.-C.; Vilar, R. Femtosecond laser surface texturing of titanium as a method to reduce the adhesion of Staphylococcus aureus and biofilm formation. *Appl. Surf. Sci.* **2016**, *360*, 485–493. [CrossRef]

16. Chan, C.W.; Carson, L.; Smith, G.C.; Morelli, A.; Lee, S. Enhancing the antibacterial performance of orthopaedic implant materials by fibre laser surface engineering. *Appl. Surf. Sci.* **2017**, *404*, 67–81. [CrossRef]

17. Chan, C.-W.; Lee, S.; Smith, G.; Sarri, G.; Ng, C.-H.; Sharba, A.; Man, H.-C. Enhancement of wear and corrosion resistance of beta titanium alloy by laser gas alloying with nitrogen. *Appl. Surf. Sci.* **2016**, *367*, 80–90. [CrossRef]

18. Faeda, R.S.; Tavares, H.S.; Sartori, R.; Guastaldi, A.C.; Marcantonio, E.J. Evaluation of titanium implants with surface modification by laser beam: Biomechanical study in rabbit tibias. *Braz. Oral Res.* **2009**, *23*, 137–143. [CrossRef]

19. Cunha, A.; Zouani, O.F.; Plawinski, L.; Botelho do Rego, A.M.; Almeida, A.; Vilar, R.; Durrieu, M.C. Human mesenchymal stem cell behavior on femtosecond laser-textured Ti-6Al-4V surfaces. *Nanomedicine* **2015**, *10*, 725–739. [CrossRef]

20. Zhou, J.; Sun, Y.; Huang, S.; Sheng, J.; Li, J.; Agyenim-Boateng, E. Effect of laser peening on friction and wear behavior of medical Ti6Al4V alloy. *Opt. Laser Technol.* **2019**, *109*, 263–269. [CrossRef]

21. Folkes, J.A. Laser Surface Melting and Alloying of Titanium Alloys. Ph.D. Thesis, University of London, London, UK, 1986.

22. Hussein, H.T.; Kadhim, A.; Al-Amiery, A.A.; Kadhum, A.A.H.; Mohamad, A.B. Enhancement of the Wear Resistance and Microhardness of Aluminum Alloy by Nd:YaG Laser Treatment. *Sci. World J.* **2014**, *2014*, 1–5. [CrossRef]

23. Mudali, U.K.; Pujar, M.G.; Dayal, R.K. Effects of Laser Surface Melting on the Pitting Resistance of Sensitized Nitrogen-Bearing Type 316L Stainless Steel. *J. Mater. Eng. Perform.* **1997**, *7*, 214–220. [CrossRef]

24. Langlade, C.; Vannes, A.B.; Krafft, J.M.; Martin, J.R. Surface modification and tribological behaviour of titanium and titanium alloys after YAG-laser treatments. *Surf. Coat. Technol.* **1998**, *100*, 383–387. [CrossRef]

25. Sun, Z.; Annergren, I.; Pan, D.; Mai, T.A. Effect of laser surface remelting on the corrosion behavior of commercially pure titanium sheet. *Mater. Sci. Eng. A* **2003**, *345*, 293–300. [CrossRef]

26. National Joint Registry. *National Joint Registry for England, Wales and Northern Ireland 11th Annual Report 2014*; Pad Creative Ltd.: London, UK, 2014.

27. National Joint Registry. *National Joint Registry for England, Wales and Northern Ireland 12th Annual Report 2015*; Pad Creative Ltd.: London, UK, 2015.

28. National Joint Registry. *National Joint Registry for England, Wales and Northern Ireland 13th Annual Report 2016*; Pad Creative Ltd.: London, UK, 2016.

29. National Joint Registry. *National Joint Registry for England, Wales and Northern Ireland 15th Annual Report 2018*; Pad Creative Ltd.: London, UK, 2018.

30. Evans, J.T.; Evans, J.P.; Walker, R.W.; Blom, A.W.; Whitehouse, M.R.; Sayers, A. How long does a hip replacement last? A systematic review and meta-analysis of case series and national registry reports with more than 15 years of follow-up. *Lancet* **2019**, *393*, 647–654. [CrossRef]

31. Chesmel, K.D.; Clark, C.C.; Brighton, C.T.; Black, J. Cellular responses to chemical and morphologic aspects of biomaterial surfaces. II. The biosynthetic and migratory response of bone cell populations. *J. Biomed. Mater. Res.* **1995**, *29*, 1101–1110. [CrossRef]

32. Leyens, C.; Peters, M. *Titanium and Titanium Alloys: Fundamentals and Applications*; Wiley-VCH: Hoboken, NJ, USA, 2003.

33. Meirelles, L.; Arvidsson, A.; Albrektsson, T.; Wennerberg, A. Increased bone formation to unstable nano rough titanium implants. *Clin. Oral Implant. Res.* **2007**, *18*, 326–332. [CrossRef] [PubMed]

34. Brown, M.S.; Arnold, C.B. *Laser Precision Microfabrication*; Sugioka, K., Meunier, M., Pique, A., Eds.; Springer: Berlin, Germany, 2010; pp. 91–120.

35. Fotovvati, B.; Wayne, S.F.; Lewis, G.; Asadi, E. A Review on Melt-Pool Characteristics in Laser Welding of Metals. *Adv. Mater. Sci. Eng.* **2018**, *2018*, 1–18. [CrossRef]

36. Chan, C.W.; Carson, L.; Smith, G.C. Fibre laser treatment of martensitic NiTi alloys for load-bearing implant applications: Effects of surface chemistry on inhibiting Staphylococcus aureus biofilm formation. *Surf. Coat. Technol.* **2018**, *349*, 488–502. [CrossRef]

37. György, E.; del Pino, A.P.; Serra, P.; Morenza, J.L. Microcolumn development on titanium by multipulse laser irradiation in nitrogen. *J. Mater. Res.* **2003**, *18*, 2228–2234. [CrossRef]

38. Picasso, M.; Hoadley, A.F.A. Finite element simulation of laser surface treatments including convection in the melt pool. *Int. J. Numer. Methods Heat Fluid Flow* **1994**, *4*, 61–83. [CrossRef]

39. Teughels, W.; Assche, N.; Van Sliepen, I.; Quirynen, M.; Van Assche, N. Effect of material characteristics and/or surface topography on biofilm development. *Clin. Oral Implant. Res.* **2006**, *17*, 68–81. [CrossRef] [PubMed]

40. Griepentrog, M.; Griepentrog, M.; Haustein, I.; Müller, W.D.; Lange, K.P.; Briedigkeit, H.; Göbel, U.B. Plaque formation on surface modified dental implants. *Clin. Oral Implant. Res.* **2001**, *12*, 543–551.

41. Brunello, G.; Brun, P.; Gardin, C.; Ferroni, L.; Bressan, E.; Meneghello, R.; Zavan, B.; Sivolella, S. Biocompatibility and antibacterial properties of zirconium nitride coating on titanium abutments: An in vitro study. *PLoS ONE* **2018**, *13*, e0199591. [CrossRef] [PubMed]

42. Badrinarayanan, S.; Sinha, S.; Mandale, A.B. XPS studies of nitrogen ion implanted zirconium and titanium. *J. Electron. Spectrosc. Relat. Phenom.* **1989**, *49*, 303–309. [CrossRef]

43. Gotman, I. Characteristics of Metals Used in Implants. *J. Endourol.* **1997**, *11*, 383–389. [CrossRef] [PubMed]

44. Di Silvio, L.; Jayakumar, P. *Cellular Response to Biomaterials*; Woodhead Publishing Limited: Cambridge, UK, 2009; pp. 313–343.

45. Gittens, R.; McLachlan, T.; Olivares-Navarrete, R.; Cai, Y.; Berner, S.; Tannenbaum, R.; Schwartz, Z.; Sandhage, K.H.; Boyan, B.D. The effects of combined micron-/submicron-scale surface roughness and nanoscale features on cell proliferation and differentiation. *Biomaterials* **2011**, *32*, 3395–3403. [CrossRef] [PubMed]

46. Weiss, P. Experiments on cell and axon orientation in vitro: The role of colloidal exudates in tissue organization. *J. Exp. Zool. Part A* **1945**, *100*, 353–386. [CrossRef]

47. Loye, A.M.; Kinser, E.R.; Bensouda, S.; Shayan, M.; Davis, R.; Wang, R.; Chen, Z.; Schwarz, U.D.; Schroers, J.; Kyriakides, T.R. Regulation of Mesenchymal Stem Cell Differentiation by Nanopatterning of Bulk Metallic Glass. *Sci. Rep.* **2018**, *8*, 1–11. [CrossRef] [PubMed]

48. Martínez, E.; Engel, E.; Planell, J.; Samitier, J. Effects of artificial micro- and nano-structured surfaces on cell behaviour. *Ann. Anat.* **2009**, *191*, 126–135. [CrossRef]

49. Cunha, A. *Multiscale Femtosecond Laser Surface Texturing of Titanium and Titanium Alloys for Dental and Orthopaedic Implants*; Universite de Bordeaux: Bordeaux, France, 2015.

50. Harrison, R.G. The cultivation of tissues in extraneous media as a method of morpho-genetic study. *Anat. Rec.* **1912**, *6*, 181–193. [CrossRef]

51. Curtis, A.; Wilkinson, C. Topographical control of cells. *Biomaterials* **1998**, *18*, 1573–1583. [CrossRef]

52. Kilian, K.; Bugarija, B.; Lahn, B.T.; Mrksich, M. Geometric cues for directing the differentiation of mesenchymal stem cells. *Proc. Natl. Acad. Sci. USA* **2010**, *107*, 4872–4877. [CrossRef] [PubMed]

53. Jimenez-Vergara, A.C. Refined assessment of the impact of cell shape on human mesenchymal stem cell differentiation in 3D contexts. *Acta Biomater.* **2019**. [CrossRef] [PubMed]

54. Mangano, C.; Piattelli, A.; D'avila, S.; Iezzi, G.; Mangano, F.; Onuma, T.; Shibli, J.A. Early Human Bone Response to Laser Metal Sintering Surface Topography: A Histologic Report. *J. Oral Implantol.* **2010**, *36*, 91–96. [CrossRef] [PubMed]

55. Helal, M.E.; Gad, E.; Helal, M.; Zaghlool, M. Effect of different titanium laser surface treatments on osseointegration. *Int. J. Acad. Res.* **2010**, *2*, 138–144.

56. Charles, P.D.; Anandapandian, P.A.; Samuel, S. Osteogenic potential of laser modified and conditioned titanium zirconium surfaces. *J. Indian Prosthodont. Soc.* **2016**, *16*, 253–258.

57. Chu, S.-F.; Huang, M.-T.; Ou, K.L.; Sugiatno, E.; Cheng, H.-Y.; Huang, Y.-H.; Chui, W.-T.; Liou, T.-H. Enhanced biocompatible and hemocompatible nano/micro porous surface as a biological scaffold for functionalizational and biointegrated implants. *J. Alloy. Compd.* **2016**, *684*, 726–732. [CrossRef]

58. Puckett, S.D.; Taylor, E.; Raimondo, T.; Webster, T.J. The relationship between the nanostructure of titanium surfaces and bacterial attachment. *Biomaterials* **2010**, *31*, 706–713. [CrossRef]

59. Chan, C.W.; Hussain, I.; Waugh, D.G.; Lawrence, J.; Man, H.C. Effect of laser treatment on the attachment and viability of mesenchymal stem cell responses on shape memory NiTi alloy. *Mater. Sci. Eng. C. Mater. Biol. Appl.* **2014**, *42*, 254–263. [CrossRef]

60. Feng, G.; Cheng, F.; Wang, S.-Y.; Borca-Tasciuc, D.A.; Worobo, R.W.; Moraru, C.I. Bacterial attachment and biofilm formation on surfaces are reduced by small-diameter nanoscale pores: How small is small enough? *Biofilms Microbiomes* **2015**, *1*, 1–9. [CrossRef]

Article

Proactive Release of Antimicrobial Essential Oil from a "Smart" Cotton Fabric

Danaja Štular [1], Matic Šobak [2], Mohor Mihelčič [2], Ervin Šest [2], Ilija German Ilić [3], Ivan Jerman [2], Barbara Simončič [1] and Brigita Tomšič [1,*]

[1] Department of Textiles, Graphic Arts and Design, Faculty of Natural Sciences and Engineering, University of Ljubljana, Aškerčeva 12, 1000 Ljubljana, Slovenia; danaja.stular@ntf.uni-lj.si (D.Š.); barbara.simoncic@ntf.uni-lj.si (B.S.)

[2] National Institute of Chemistry, Hajdrihova 19, 1000 Ljubljana, Slovenia; matic.sobak@ki.si (M.Š.); mohor.mihelcic@ki.si (M.M.); ervin.sest@ki.si (E.Š.); ivan.jerman@ki.si (I.J.)

[3] The Department of Pharmaceutical Technology, Faculty of Pharmacy, University of Ljubljana, Aškerčeva 7, 1000 Ljubljana, Slovenia; ilija.german.ilic@ffa.uni-lj.si

[*] Correspondence: brigita.tomsic@ntf.uni-lj.si; Tel.: +386-1-20-03-233

Received: 25 March 2019; Accepted: 7 April 2019; Published: 10 April 2019

Abstract: Two temperature and pH responsive submicron hydrogels based on poly(*N*-methylenebisacrylamide), chitosan and β-cyclodextrines (PNCS/CD hydrogel) with varying poly(*N*-isopropylacrylamide) to chitosan ratios were synthesized according to a simplified procedure, reflecting improved stimuli responsive properties and excellent bio-barrier properties, granted by incorporated chitosan. Hydrogels were applied to cotton-cellulose fabric as active coatings. Subsequently, antimicrobially active savory essential oil (EO) was embedded into the hydrogels in order to develop temperature- and pH-responsive cotton-cellulose fabric with double antimicrobial activity, i.e., bio-barrier formation of chitosan along with the proactive release of savory EO at predetermined conditions. The influence of the hydrogels chemical composition on stimuli responsive and antibacterial properties were assessed. Both PNCS/CD hydrogels showed stimuli responsiveness along with controlled release of savory EO. The chemical composition of the hydrogels strongly influenced the size of the hydrogel particles, their temperature and pH responsiveness, and the bio-barrier forming activity. The increased concentration of chitosan resulted in superior overall stimuli responsiveness and excellent synergy between the antimicrobial activities of the hydrogel and released savory EO.

Keywords: antimicrobial activity; chitosan; essential oils; smart textiles; stimuli-responsive hydrogel; β-cyclodextrin

---

1. Introduction

Essential oils (EOs), as one of the important representatives of the natural antimicrobial agents, play a vital role in the textile industry. Namely, they are used to produce biocompatible, environmentally friendly and non-toxic high value-added products with antimicrobial, anticancer, antioxidant and aromatherapeutic properties, intended for the use as cosmetotextiles or medical textiles [1–4]. However, the application of EOs in textiles is limited due to their sensitivity to various environmental conditions. Namely, some volatile compounds are lost if EOs are stored at high temperatures, while other components become unstable due to variations in the pH or are likely to undergo the oxidation process and generation of free radicals when the EOs are exposed to the combination of oxygen and light [5].

One possible approach for the incorporation of EOs into textile materials is their embedment into stimuli-responsive hydrogels. In this manner, the EOs are not only protected against the environment but are also proactively released from the hydrogel matrix only at predetermined conditions. Such controlled release of entrapped substances is based on the ability of the hydrogel particles to reversibly swell or de-swell due to the presence or absence of different external stimuli [6]. However, two important issues should be considered when embedding the EOs as an active substance into a stimuli-responsive hydrogel matrix. The first is the hydrophobic nature of EOs because it is known that water is a driving force of hydrogel responsiveness and the second is the concentration-dependent antibacterial activity of the released EOs, and thus the lack of antimicrobial efficiency when the hydrogel is in its swollen phase. In addition, the hydrogel matrix acts as a reservoir of the EOs, meaning that there is limit amount of EOs that can be embedded, which lead to the eventual depletion of the EOs during the product use. The first issue can be successfully resolved by introducing β-cyclodextrines (β-CD) into the hydrogel structure. Specifically, because β-CD consist of a hydrophobic cavity and a hydrophilic surface, they can form an inclusion complex with appropriately sized molecules [7,8]. Accordingly, a successful incorporation of β-CD into poly(*N*-vinylcaprolactam)-based hydrogel was demonstrated by Kettle et al. by using a two-step synthesis procedure that included the formation of β-CD acrylates in the first step, followed by a further copolymerization of the hydrogel in the second step [9,10]. However, the two-step synthesis method of the incorporation of reactive β-CD functionalized by acrylic groups into the hydrogel matrix is not only time-consuming but is also rather complex, thus the preparation of cyclodextrin acrylates proceeds according to the sophisticated Schlenk technique where the utilization of a less environmentally susceptible organic solvent is necessary. Yi and co-workers have recently reported on the effective, ecologically more susceptible incorporation of β-CD into poly-(*N*-isopropylacrylamide) (poly-NiPAAm) based hydrogel via a one-step method directly during the hydrogel synthesis process in order to achieve temperature responsiveness [11]. In this study it was also shown that the presence of β-CD was beneficial for the increased thermosensitivity of the poly-NiPAAm, which was ascribed to the formation of β-CD-poly-NiPAAm conjugation.

The second issue regarding the concentration-dependent antibacterial activity of the released EOs within the hydrogel matrix, remains a rather unexplored topic. Accordingly, one possible approach to extend the life span of the antimicrobial activity, but preserve the other beneficial properties of the EOs, is to combine EOs with bio-barrier forming agents, resulting in the formation of dual antibacterial activity. Namely, the bio-barrier forming agents do not leach from the fibers, but create a biological obstacle for the microorganisms that comes in direct with the fibers, meaning that their concentration does not change with time. In this manner, antimicrobial activity of the functionalized fibers could be obtained even in conditions when the beneficial effect of the EOs is minimized, i.e., the hydrophilic, swollen phase of the hydrogel particles, or when the concentration of EOs on the surface of the fibers would decrease below the limit of efficiency.

Among the stimuli-responsive hydrogels, the poly-NiPAAm/chitosan based hydrogel (PNCS hydrogel) is particularly important owing to its response to both temperature and pH in physiological range [12]. Accordingly, PNCS hydrogel swells at temperatures below 32 °C (i.e., below the lower critical solution temperature (LCST) of poly-NiPAAm) and/or pH below 6.5 (i.e., below the pK$_a$ of chitosan), while it undergoes the transition to the de-swollen, collapsed phase when the temperature and/or pH rise above the aforementioned values. Bearing in mind that pH-responsive chitosan is a well-known biocompatible bio-barrier forming antimicrobial agent [13–15], dual antibacterial activity can be achieved without the use of additional bio-barrier forming agents, but rather by the simple increase of the chitosan within the PNCS hydrogel. The proposed mechanism of the present study, showing dual antimicrobial activity, is schematically presented in Figure 1. Moreover, it can be also predicted that the increase in the chitosan concentration will not compromise dual temperature and pH responsiveness of functionalized PNCS/CD hydrogel, as incorporation of β-CD increases the thermosensitivity of poly-NiPAAm.

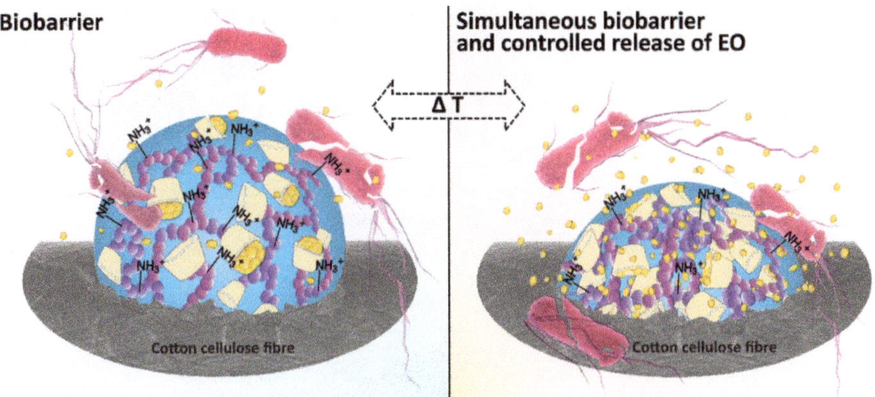

Figure 1. Proposed dual antimicrobial activity of the poly-(*N*-isopropylacrylamide) (poly-NiPAAm)/chitosan based hydrogel (PNCS/CD) hydrogel in correlation with the changes in temperature of the environment.

To the best of our knowledge, there have been no studies about the mutual influence between PNCS/CD hydrogel structure and EOs, thus based on the above consideration, a comprehensive study was performed. Accordingly, this paper focuses on the synthesis PNCS/CD submicron hydrogels with two different poly-NiPAAm to chitosan ratios of 7:1 and 4:1 using a modified in situ synthesis procedure. Subsequently, the PNCS/CD hydrogels were applied to the cotton-cellulose fabric as a thin layer surface modifying system and were functionalized with savory EO, which showed to provide excellent antibacterial activity, without influencing the stimuli responsiveness of the PNCS hydrogel [16]. In this manner, we aimed to study (i) the influence of the β-CD incorporation on the temperature and/or pH responsiveness of the PNCS/CD hydrogel, (ii) the influence of the increased chitosan concentration on the temperature responsiveness of the PNCS/CD hydrogel and (iii) the possible achievement of synergistic dual antimicrobial activity derived by the proactive release of savory EO and bio-barrier-forming chitosan.

2. Materials and Methods

2.1. Materials

In this study, alkaline-scoured, bleached, and mercerized 100% cotton-cellulose plain weave fabric with a mass area of 125 g/m^2 (warp density: 38 threads/cm; weft density: 28 threads/cm) was purchased from Tekstina d.d., Ajdovščina, Slovenia. Savory (*Satureja montana* L.) essential oil (Florihana Distillerie) was purchased from Magnolija, Nina Medved IC, Ljubljana, Slovenia. The chemical composition of savory EO provided by Florihana Destillerie (Caussols, France.) is presented in Supplementary 1 (Table S1). Tween 20 (Sigma-Aldrich, St. Louis, MO, USA) was used as surfactant. For the preparation of the PNCS/CD hydrogel, chitosan (Chitoclear, Primex, Siglufjordur, Iceland; DD = 95%; η = 159 mPa), glacial acetic acid (Sigma-Aldrich), *N*-isopropylacrylamide (NiPAAm) (Sigma-Aldrich), *N,N*-methylenebisacrylamide (MBA) (Sigma-Aldrich), ammonium persulfate (APS) (Sigma-Aldrich), β-cyclodextrine (β-CD) (Sigma-Aldrich) and highly pure sodium dodecyl sulfate (SDS, GE Healthcare Life Sciences, São Paulo, Brazil) were used.

2.2. Synthesis of the PNCS/CD Hydrogels

The PNCS/CD hydrogels were synthesized according to the combined and modified procedures firstly reported by Lee [17] and Yi et al. [11]. Briefly, the appropriate concentration of chitosan was dissolved in a mixture of 300 mL of distilled water and 3 mL of glacial acetic acid, placed in a flask and degassed with nitrogen for 30 min. Next, NiPAAm, β-CD, 10 wt.% SDS solution in water and MBA were added while intensely stirring the mixture at 320 rpm and heating the reaction mixture to 50 °C. After 40 min, APS was added to initiate the polymerization. The mixture turned turbid in approximately 10 min, verifying that the polymerization was proceeding. The reaction proceeded in a nitrogen atmosphere at 50 °C for 3 h. Subsequently, the PNCS/CD hydrogels reaction mixtures were dialyzed against bidistilled water using a Spectra/Por 4 membrane (Fisher Scientific, Pittsburgh, PA, USA) for 10 days to remove impurities and unreacted monomers and the water was changed daily. The studied PNCS/CD hydrogels have different ratios of poly-NiPAAm and chitosan (PN:CS). The hydrogel sample codes, and the used reagents and their concentrations are presented in Table 1.

Table 1. Synthesis reagents used for the preparation of the PNCS and PNCS/CD hydrogels.

Hydrogel Code	PN:CS Ratio	Reagents (g)					
		NiPAAm	Chitosan	β-CD	SDS Solution	MBA	APS
PNCS/CD M1	7:1	7.00	1.00	14.00	1.50	0.21	0.90
PNCS/CD M2	4:1	7.00	1.75	14.00	1.55	0.22	0.93

2.3. Hydrogel Dispersion Characterization

The morphology of the hydrogel particles in the dispersions was determined by scanning electron microscopy at an accelerating voltage of 1 kV using an FE-SEM Zeiss SUPRA 35 VP instrument (Oberkochen, Germany). The hydrogel dispersions were freeze-dried, mixed with water, applied on the holder and dried. Prior to the observation, the samples were coated with Pt.

The particle sizes of the dispersed PNCS and PNCS/CD hydrogels as a function of the temperature were determined by dynamic light scattering (DLS) using a Zetasizer NanoS (Malvern, UK) equipped with 4 mW He-Ne laser operating at the wavelength of 633 nm and an avalanche photodiode detector. Scattering light was detected at an angle of 173°. Hydrogel was diluted 50 times so that 60 μL of the sample was used for each measurement, and the results represent an average of three values. The particle size was determined at pH 6.5 and temperatures in the 20–40 °C range at intervals of 5 °C.

Thermal analysis was performed by differential scanning calorimetry (DSC) carried out using a DSC 1 calorimeter (Mettler Toledo, Columbus, OH, USA) that was calibrated with indium. Approximately 8–10.5 mg of a hydrogel sample was accurately weighted into a 40 μL aluminum pan that was then hermetically sealed with an aluminum lid. The thermal analyses were performed at the temperatures varying from 5 to 50 °C at a heating rate of 1.5 K/min using a nitrogen atmosphere. One cycle of heating and cooling was performed. Normalized thermograms were evaluated using the STARe v9.30 software (Mettler Toledo, Columbus, OH, USA), where the point of the endothermic peak was used to indicate the LCST of the hydrogels.

2.4. Application of the Hydrogels onto the Cotton-Cellulose Fabric

PNCS and PNCS/CD hydrogels were applied to a cotton-cellulose fabric using a pad-dry method, which included: immersion of the fabric samples into a hydrogel dispersion, wringing by a two-roller foulard (Mathis, Oberhasli, Switzerland) with a wet pick-up (WP) of 80% ± 5% and 5 min drying of the samples at 80 °C, using laboratory dryer (Mathis, Oberhasli, Switzerland). The pressure on the foulard varied based on the WP that was calculated for each sample according to [18]:

$$WP = \left(\frac{m_w - m_d}{m_d}\right) \times 100 \ [\%] \tag{1}$$

where m_d (g) is the weight of the dry, untreated sample and m_w (g) represents the weight of the immersed sample after wringing.

2.5. Essential Oil Emulsion Preparation

The appropriate concentration of EO in emulsion was determined according to the disc diffusion test, which provided the information regarding the antimicrobial activity of EO in concentrations from 1% to 5%. Results are presented in Supplementary (Figure S1). Based on the results, 5% savory EO emulsion was prepared for the embedment into the studied PNCS/CD hydrogels. Firstly, 10 g of Tween 20 was diluted in 85 g of distilled water and mixed for 20 min at room temperature with the speed of 100 rpm. Then, the stirring speed was set to 1500 rpm, while 5 g of savory EO was added dropwise into the mixture. The mixing proceeded for 4 h and was followed by ultra-sonication carried out with a lab homogenizer (Hielscher Ultrasonics GmbH, Brandenburg, Germany) for 5 min using 0.5 ultrasonic impulses.

2.6. Embedment of the EO Emulsion

Immediately after drying the PNCS/CD coated fabric samples were immersed into a savory EO emulsion, which was previously stored in a refrigerator at 8 °C in order to fasten the swelling of the shrunken hydrogel particles. The wet samples were pressed with the foulard, to achieve WP of 80% ± 5%. Subsequently, the samples were immersed in a bath containing distilled water:ethanol (60:40) mixture at a goods to liquid ratio of 1:50 in order to remove the excess of the savory EO emulsion.

To characterize the amount of the functional finishing present on the fabric samples, Add-on was determined for each sample. The samples were placed in the moisture analyzer MLB-C (Kern & SOHN GmbH, Balingen, Germany) and dried to their constant weight. The Add-on was calculated according to following equation [18]:

$$\text{Add} - \text{on} = \left(\frac{m_f - m_{UN}}{m_{UN}} \right) \times 100 \; [\%] \tag{2}$$

where m_f (g) is the average of the dry weights of the functionalized sample, and m_{UN} (g) represents an average of dry weights of the untreated sample. Each value represents an average of 10 measurements.

The sample codes and Add-on values of all of the studied samples are presented in Table 2.

Table 2. Sample codes according to the chemical modification of the functionalized cotton-cellulose fabric and corresponding Add-on values.

Sample Code	Description of Chemical Modification	Add-on (%)
CO_UN	/[a]	/[a]
CO_M1	Cotton fabric treated with PNCS/CD hydrogel M1 (PN:CS = 7:1)	7.86
CO_M2	Cotton fabric treated with PNCS/CD hydrogel M2 (PN:CS = 4:1)	8.89
CO_M1+S	Cotton fabric treated with PNCS/CD hydrogel M1 and savory EO emulsion	18.19
CO_M2+S	Cotton fabric treated with PNCS/CD hydrogel M2 and savory EO emulsion	18.88

[a] No finishing.

2.7. Analysis and Measurements

2.7.1. Scanning Electron Microscopy (SEM)

The morphology of the samples was determined by scanning electron microscopy at an accelerating voltage of 1 kV using an FE-SEM Zeiss SUPRA 35 VP instrument (Oberkochen, Germany). Prior to the observation, the samples were coated with Pt.

2.7.2. Fourier Transform Infrared (FT-IR) Spectroscopy

Fourier-transform infrared (FT-IR) spectra were obtained using a Spectrum GX I spectrophotometer (Perkin Elmer, Waltham, MA, USA) equipped with an attenuated total reflection (ATR) cell and a diamond crystal (n = 2.0). The spectra were recorded over the range 4000–600 cm^{-1}, with a resolution of 4 cm^{-1} and averaged over 32 spectra.

2.7.3. Moisture Content (MC)

Temperature responsiveness of the samples was assessed by measuring moisture content according to the AATCC D629—15 Standard Test Methods for Quantitative Analysis of Textiles [19]. For this purpose, a moisture analyzer MLB-C (Kern & SOHN GmbH, Balingen, Germany) was used. Studied samples were pre-conditioned at 65% $\pm$ 2% relative humidity at 20 and 40 °C for 24 h, before they were put in a moisture analyzer and dried at 60 °C until the constant mass. Moisture content (MC) was determined by Equation (3) and each value represents an average of 10 measurements.

$$\mathrm{MC} = \left(\frac{m_0 - m_\mathrm{f}}{m_0} \right) \times 100 \; [\%] \tag{3}$$

where m_0 denote the initial mass of the pre-conditioned sample and m_f represents the final mass of the sample after drying. MC was reported as mean values of 10 measurements.

Based on the results of the average MC value, the contribution (C_{MC}) of both hydrogels on the studied samples was calculated by the following equation:

$$C_{MC} = \left(\frac{\mathrm{MC_F} - \mathrm{MC_{UN}}}{\mathrm{MC_{UN}}} \right) \times 100 \; [\%] \tag{4}$$

where $\mathrm{MC_F}$ is the average moisture content value of the studied functionalized sample, and $\mathrm{MC_{UN}}$ is the average moisture content value of the untreated sample, determined under the same conditions.

2.7.4. Water Uptake (WU)

In order to evaluate pH responsiveness of the functionalized samples, water uptake was determined [20]. Firstly, the samples were dried to their constant weight using a moisture analyzer MLB-C (Kern & SOHN GmbH, Balingen, Germany) and their dry weight was determined. Next, they were immersed into the buffer solution with pH 3 for 1 h, followed by immersion for 1 h in buffer solution with pH 8 and immersion for 1 h in buffer solution with pH 3. During the analysis, the samples were taken out of the solutions and weighed every 15 min, thus water uptake (WU) was determined for each measurement according to Equation (5). Each value represents an average of 10 measurements.

$$\mathrm{WU} = \left(\frac{m_\mathrm{w} - m_0}{m_0} \right) \times 100 \; [\%] \tag{5}$$

where m_w is the weight of the sample taking up buffer solution (g) and m_0 is the dry weight of the sample (g).

All data were presented as the mean and standard deviation values. To determine the significance of the differences between studied sets of data, a t-test was performed, and the significance limit was set at $p = 0.05$.

2.7.5. Antimicrobial Activity

Antibacterial activity of studied cotton samples was estimated by determination of zone of inhibition referring to the ISO 20645 Agar diffusion plate test [21], as well as by determination of bacterial reduction according to the ISO 20743 Absorption method [22]. As test organisms, Gram-negative bacteria *Escherichia coli* (ATCC 25922) and Gram-positive bacteria *Staphylococcus aureus* (ATCC 6538) were selected.

According to the ISO 20645 Agar diffusion plate test, assessment of antibacterial activity was based on the absence or presence of an inhibition zone, which was calculated from:

$$H = \frac{D - d}{2} \; (mm) \tag{6}$$

where H is the inhibition zone in mm, D is the total diameter of the cotton specimen and inhibition zone in mm, and d is the diameter of specimen in mm. All tests were performed in duplicate. After the test, the samples and the zone of inhibition were observed with a Leica EZ4 W optical microscope (Leica, Wetzlar, Germany).

According to the ISO 20743 standard method, 1 g of sample was incubated with bacterial suspension of 10^5 CFU/mL. The number of surviving colonies was determined by spreading 40 µL of suspension on an agar plate immediately after the inoculation (time 0) and after 1 h (time 1) of bacteria suspension exposure. The amount of inoculum was adapted to the size of the agar plate. Accordingly, the chosen volume was taken into account in further calculations, adapting to a volume of 1 mL.

The antibacterial activity value was calculated as follows:

$$A = (\lg C_1 - \lg C_0) - (\lg T_1 - \lg T_0) \tag{7}$$

where A is the antibacterial activity value, $\lg C_1$ is the average common logarithm for the number of bacteria obtained from control sample after 1 h of incubation, $\lg C_0$ is the average common logarithm for the number of bacteria obtained from the control sample immediately after inoculation, $\lg T_1$ is the average common logarithm for the number of bacteria obtained from tested sample after 1 h of incubation and $\lg T_0$ is the average common logarithm for the number of bacteria obtained from the tested sample immediately after inoculation. For each sample, the values represent the average of eight counts.

3. Results and Discussion

3.1. Characterization of the Hydrogels

The morphology and particle size of the studied PNCS/CD M1 and M2 hydrogels with different PN:CS ratios were determined by SEM and DLS analysis. In addition, the temperature of the phase change transition of the studied hydrogel particles was also determined. The representative SEM images of these samples are shown in Figure 2. The PNCS/CD M1 hydrogel particles were spherically shaped, but in some cases, the outline of the individual particles was blurred, implying that these particles were coated. In contrast, the spherical shape of the hydrogel particles of PNCS/CD M2 was more distinct with clearly formed individual segments. Since direct comparison between the particle sizes of the two hydrogels was not possible, the DLS method was employed.

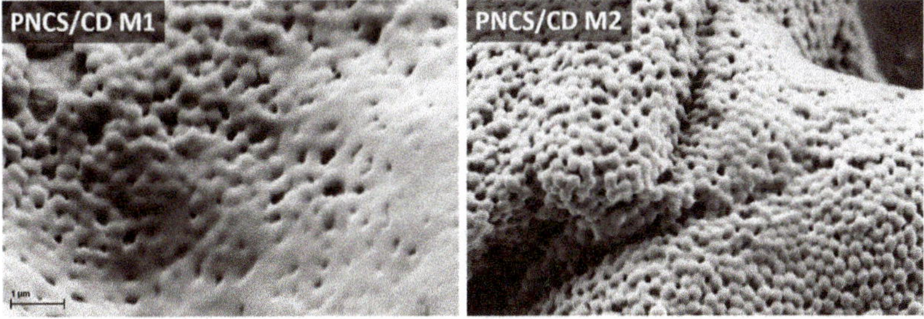

Figure 2. Scanning electron microscope images of the studied PNCS/CD hydrogels.

The results of the hydrodynamic particle size of the studied PNCS/CD hydrogels in correlation with the fluctuation in the temperature of the intermediate surroundings are presented in Figure 3. Regardless of the polymer ratio, both of the studied PNCS/CD hydrogels had submicron size particles at 20 °C, responding to the temperature increase by decreasing their hydrodynamic size. However, for the PNCS/CD M1 hydrogel, this decrease was only obtained until the temperature reached 30 °C, whereas at higher temperatures, the hydrodynamic size of the particles increased dramatically. This behavior can be ascribed to the fast agglomeration of the PNCS/CD M1 hydrogel particles at the temperatures above the LCST of poly-NiPAAm, as was confirmed by the drastic increase in the polydispersity index (PDI) values. The rapid agglomeration of the PNCS/CD M1 hydrogel particles can be attributed to the formation of an inclusion complex between the β-CD and poly-NiPAAm that enhanced the temperature responsiveness [11]. For the PNCS/CD M2 hydrogel, the agglomeration process was not observed, thus confirming that the presence of higher concentration of chitosan stabilizes the hydrogel particles during the synthesis procedure [23]. Namely, at 20 °C, when poly-NiPAAm is in its hydrophilic state and the particles are swollen, the hydrodynamic particle size of the PNCS/CD M2 hydrogel was 20% smaller than that obtained for the PNCS/CD M1 hydrogel. The hydrodynamic size was slowly decreased by gradually raising the temperature over the entire studied temperature range, showing the reduction of 66% in the particle size at the end of the experiment. Furthermore, comparison of the hydrodynamic particle size of the pure PNCS hydrogel that was determined in our previous research [24] to the hydrodynamic particle size of the PNCS/CD M1 hydrogel shows that at the same PN:CS ratio, the inclusion of β-CD into the hydrogel structure gave rise to an approximately 14% increase in the particle size. This finding unambiguously demonstrates the beneficial effect of the higher chitosan concentration in the formation of smaller and more stable particles for the PNCS/CD M2 hydrogel.

As observed from Figure 4, the LCST was found at 32.15 °C for the PNCS/CD M1 hydrogel and at 32.16 °C for the PNCS/CD M2 hydrogel. Both temperatures are in agreement with the LCST of the linear poly-NiPAAm homopolymer (LCST of approximately 32 °C), implying that neither the incorporation of β-CD nor the increased chitosan concentration showed any effect on the LCST of the studied hydrogels.

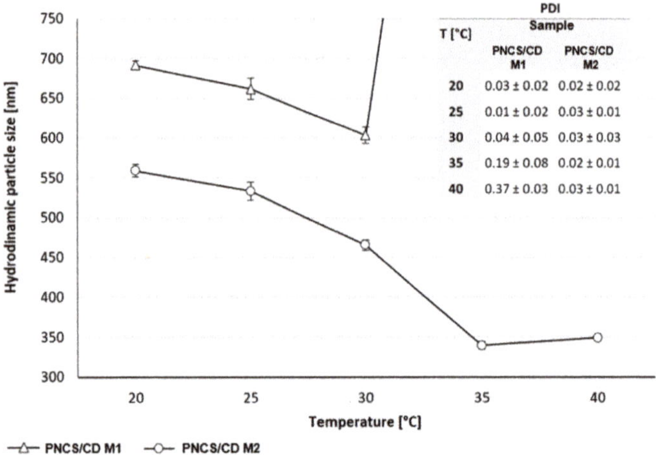

Figure 3. Hydrodynamic PNCS/CD hydrogels particle diameter in the dispersion with the corresponding polydispersity index (PDI) values.

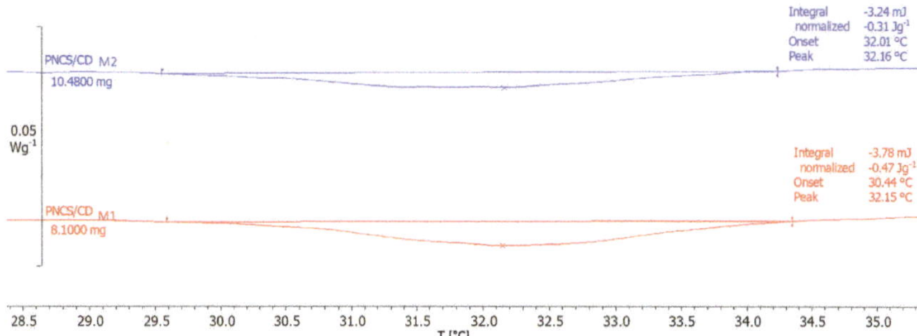

Figure 4. Differential scanning calorimetry (DSC) thermograms of the studied PNCS/CD hydrogels.

3.2. Morphological and Chemical Properties of the Functionalized Cotton-Cellulose Samples

The deposition of the stimuli responsive hydrogel altered the morphology of the cotton-cellulose fiber surface due to the presence of spherically shaped bulges (Figure 5). There is a distinct difference in the size of the microgel bulges of both hydrogels applied on cotton fabric. Namely, the hydrogel particles on CO_M1 and CO_M2 samples measured ~550 and ~350 nm in diameter, respectively. The values were smaller compared to the ones gathered by DLS (Figure 3), due to the vacuum present in SEM machine. Additionally, the water is the driving force of swelling and shrinking of the particles, therefore the particles in dispersion were able to swell freely. Nevertheless, the difference in particle size on the fiber surface correlates to their hydrodynamic size in aqueous media. After in situ embedment of the savory EO emulsion into the PNCS/CD M1 and PNCS/CD M2 hydrogel particles applied on cotton fabric (samples CO_M1+S and CO_M2+S), the volume of the hydrogel bulges visibly increased, which indicates the successful incorporation of EO into the microgel particles.

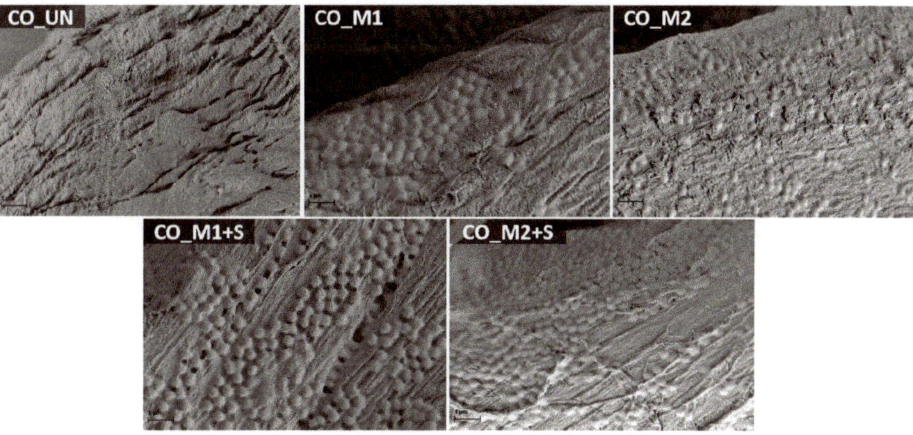

Figure 5. Scanning electron microscopy images of the untreated sample (CO_UN) and the samples finished with bare and functionalized PNCS/CD M1 and M2 hydrogels (CO_M1, CO_M2 and CO_M1+S, CO_M2+S) obtained at the magnification of 25,000 times.

FT-IR ATR spectroscopy was used to obtain information about the characteristic molecular groups and species of the studied surface modification system. To avoid the interference of the characteristic absorption bands of the cotton-cellulose fibers to the characteristic absorption bands of both PNCS/CD hydrogels and the savory EO, the FT-IR ATR analysis was first performed independently of the cotton-cellulose fibers (Figure 6). For this purpose, only the PNCS/CD M1 hydrogel was used, because it was expected that no additional information would be obtained by also analyzing the PNCS/CD

M2 hydrogel. Accordingly, the PNCS/CD M1 hydrogel was first deposited on a Si wafer, and its IR ATR spectrum was compared to that of the pure PNCS M1 hydrogel in order to confirm the inclusion of β-CD. An examination of the IR ATR spectrum shows the presence of pure PNCS hydrogel absorption bands belonging to both poly-NiPAAm and chitosan, i.e., the absorption bands at 1630 and 1530 cm^{-1} that are characteristic of the C=O stretching vibration of amide I and N–H deformation vibration of amide II; the absorption band at 3077 cm^{-1} that is due to N–H stretching; the absorption bands in the 3600–3200 cm^{-1} spectral region arising due to the N–H···OH and OH vibrations; the absorption bands at 2970, 2932 and 2875 cm^{-1} that are ascribed to the stretching vibration of CH of the N-isopropyl groups and the polymer backbone of poly-NiPAAm and chitosan; the absorption bands in the 1480–1300 cm^{-1} spectral region that are due to the CH$_3$ stretching and CH$_3$ and CH$_2$ deformation vibrations; and, finally, the absorption bands at 1170 and 1130 cm^{-1} ascribed to the polysaccharide structure of chitosan (Figure 6a) [25–31]. These absorption bands are well observed in the IR ATR spectra of the PNCS/CD M1 hydrogel (Figure 6b), where a conspicuous appearance of the new absorption bands at 1077 and 1035 cm^{-1} that are ascribed to the C–O–C, C–O and C–H vibrations indicates the successful introduction of β-CD in the PNCS hydrogel [32–34].

Second, the PNCS/CD M1 hydrogel deposited on the Si wafer was in situ functionalized by savory EO in the same manner as for cotton-cellulose fibers in order to obtain information regarding the EO entrapment into the β-CD host cavity inside the hydrogel matrix. As expected, in the corresponding spectrum (Figure 6c), the characteristic absorption bands of hydrogel and savory EO (Figure 6d) were observed, with the absorption bands of the PNCS/CD M1 hydrogel dominating. Nevertheless, the presence of the savory EO can be inferred from the strong increase in the absorption band at 3385 cm^{-1} that is ascribed to a large amount of hydroxyl groups in the water emulsion of savory EO and from the appearance of various absorption bands in the 1730–800 cm^{-1} spectral region that belong to the various compounds of the savory EO such as carvacrol, thymol, p-cymene and γ-terpinene that give rise to the absorption bands characteristic of phenols, monoterpene hydrocarbons, alcohols and ketones, sesquiterpenic hydrocarbons and alcohols, aliphatic hydrocarbons, and esters [13,35,36]. In addition, the absorption band of the amide II characteristic of the hydrogel was redshifted from 1530 to 1578 cm^{-1}, indicating a certain interaction between the hydrogel and savory EO emulsion.

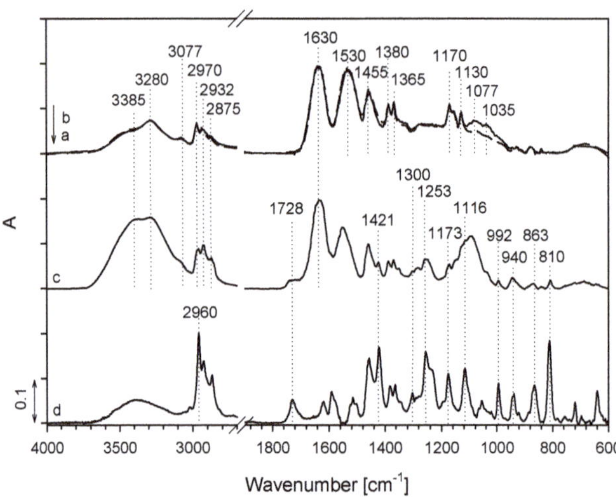

Figure 6. Infrared attenuated total reflection (IR ATR) spectra of the pure PNCS M1 hydrogel (**a**), PNCS/CD M1 hydrogel before (**b**) and after in situ functionalization by water emulsion of savory essential oil (EO) (**c**) and of pure savory EO (**d**) deposit on a Si wafer.

A comparison of the ATR IR spectra of the untreated sample with the ATR IR spectra of the functionalized cotton samples CO_M1, CO_M2 and CO_M1+S, CO_M2+S presented in Figure 7 revealed the appearance of absorption bands of amide I and amide II in the ATR IR spectra of both the CO_M1 and CO_M2 samples, thus proving the presence of the PNCS/CD M1 and PNCS/CD M2 hydrogels. The intensities of both absorption bands were slightly higher in the IR ATR spectrum of the CO_M1 sample, indicating the greater concentration of the PNCS/CD M1 hydrogel on the fibers of the CO_M1 sample. After embedding of savory EO, the IR ATR spectra of the CO_M1+S and CO_M2+S samples revealed the presence of the 1728, 1465 and 810 cm^{-1} absorption bands only. The observed absorption bands are in good agreement with those detected in the IR ATR spectrum of pure savory EO, thus clearly proving the presence of savory EO on the functionalized CO_M1+S and CO_M2+S cotton-cellulose samples.

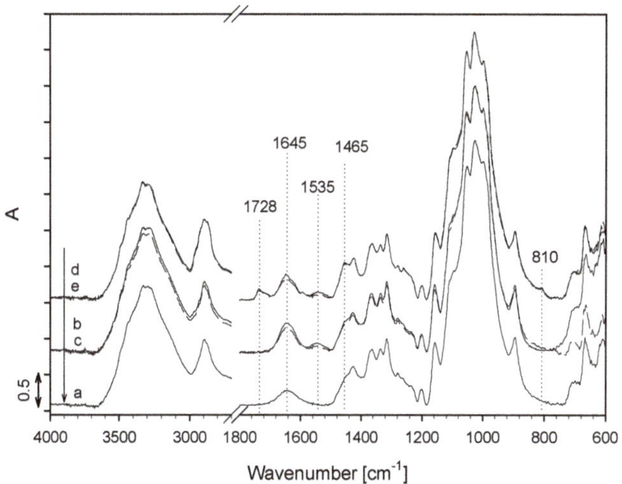

Figure 7. IR ATR spectra of the studied cotton-cellulose samples: CO_UN (**a**), CO_M1 (**b**), CO_M2 (**c**), CO_M1+S (**d**) and CO_M2+S (**e**).

3.3. Responsive Properties

Responsive properties of the functionalized samples were studied in terms of the temperature and pH responsiveness. In these studies, only the results for the CO_M1 and CO_M2 samples coated with bare PNCS/CD M1 and PNCS/CD M2 hydrogels are shown and discussed. Namely, due to the presence of volatile compounds, the presence of savory EO strongly interferes with the results obtained for the CO_M1+S and CO_M2+S samples, because a much higher mass difference between the moisturized/wet and completely dry samples was gained. This gave the misleading impression of a higher degree of responsiveness of the CO_M1+S and CO_M2+S samples compared to the CO_M1 and CO_M2 samples.

The results for the temperature responsiveness determined as the moisture content (MC) after the preconditioning of the samples at 20 or 40 °C and 65% R.H. are presented in Figure 8a. As expected, at 20 °C, both of the studied hydrogels increased the MC of the CO_M1 and CO_M2 samples compared to the CO_UN sample. This effect was ascribed to their hydrophilic character at the given conditions, resulting in comparable MC values of the functionalized samples that were approximately 13% higher than that obtained for the CO_UN sample. By contrast, the rise in the temperature triggered the transition of the PNCS/CD M1 and M2 hydrogels from the hydrophilic to the hydrophobic state, resulting in the expulsion of water from their structures. Therefore, at 40 °C, a reduction in the MC values was observed. In addition, at these conditions, both samples showed comparable MC values that were similar to that obtained for the untreated CO_UN sample. Therefore, no useful information could be extracted from the direct comparison of the samples, and thus, the contribution of the surface modifying system to the MC (C_{MC}) was also determined (Figure 8b). The results demonstrate the superior moisture management of the PNCS/CD M2 hydrogel, which is in agreement with the findings of the DLS analysis (Figure 3), i.e., the beneficial effect of the higher chitosan concentration for the stabilization and reduction in the size of the PNCS/CD M2 hydrogel particles. Specifically, smaller particles have greater specific surface area [6] and therefore exhibit a greater degree of swelling. On the other hand, regarding the influence of the β-CD inclusion on the temperature responsiveness of the PNCS hydrogel, it must be mentioned that β-CD hindered the moisture management properties compared to the pure PNCS hydrogel, which was well-studied and reported in our previous study [24]. At this point, it is important to note that the impaired moister management of the PNCS/CD hydrogel was reflected in its strong decrease of MC at 40 °C, showing that the presence of β-CD strongly affected moisture release. The most likely reason for this is the increased overall polymers' concentration in the PNCS/CD hydrogels due to the inclusion of the β-CD. Hence, it is known that the release of the moisture from the hydrogel depends on its colloidal properties including high surface area, colloidal stability and control over particle size, as well as their internal network structure characterized by the interaction with embedded compounds, mesh size and polymer volume fraction [6]. In this manner, the higher overall concentration of the polymer networks affects the porosity and tortuosity of the PNCS/CD hydrogels compared to the pure PNCS hydrogel, thus limiting the moisture release at the temperatures inducing the de-swelling of the particles.

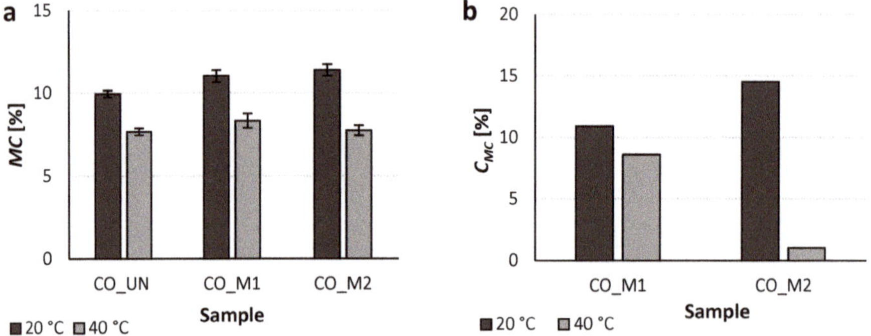

Figure 8. (**a**) Moisture content (MC) of the studied samples, determined after preconditioning at 20 or 40 °C and 65% ± 2% relative humidity for 24 h; (**b**) contribution of the hydrogels (C_{MC}) calculated from the MC values.

To investigate the pH responsiveness of the functionalized CO_M1 and CO_M2 samples, the water uptake (WU) after their sequential immersion in buffer solutions with pH 3 and 8 was determined. In addition, *p*-values were calculated to ensure that the obtained results were statistically significant (Figure 9).

The CO_UN sample showed no statistically significant difference when transferred from pH 3 to 8 and back, so the greatest increase in WU for this sample occurred within the first 15 min of immersion in pH 8, which can be ascribed to the nature of the cotton-cellulose fibers and their swelling in alkaline conditions [37]. Both the CO_M1 and CO_M2 samples respond to the pH variations in a similar manner. Namely, the WU dropped when the pH of the environment rose from 3 to 8, as the amino groups of chitosan deprotonated and the molecule shrank and expelled the aqueous media. However, when both samples were returned to a buffer solution with acidic pH, the WU increased again, as chitosan was protonated and its molecule extended due to water absorption. As expected, the CO_M2 sample showed the highest WU values due to its highest concentration of the pH-responsive chitosan in the PNCS/CD M2 hydrogel. The differences in WU may seem to be minor, but it should be pointed out that the experiment was conducted at room temperature, i.e., when poly-NiPAAm was hydrophilic, so the values only represent the pH responsiveness caused by chitosan that was present in much lower concentration as the temperature-responsive polymer. Nevertheless, the *p*-values confirm that the differences in the WU values obtained for the CO_M1 and CO_M2 samples were smaller than 0.001 and therefore can be considered to be statistically significant.

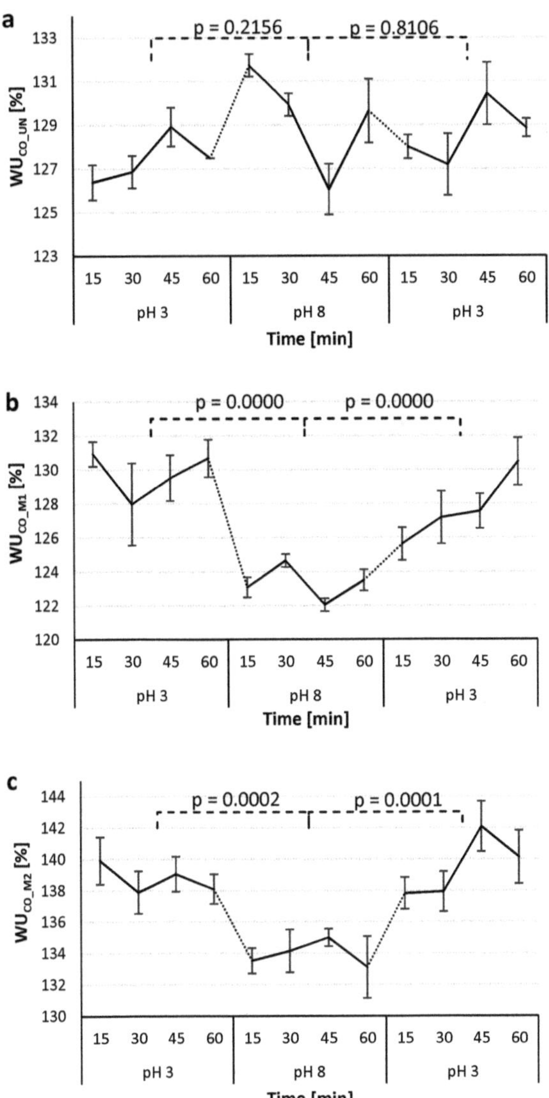

Figure 9. Water uptake (WU) of the studied CO_UN (**a**), CO_M1 (**b**) and CO_M2 (**c**) samples determined each 15 min during alternating 1 h immersion of the samples in buffer solutions of pH 3 and pH 8.

3.4. Antimicrobial Activity

Antibacterial activity was determined in order to verify the controlled release of the savory EO emulsion from the PNCS/CD M1 or PNCS/CD M2 hydrogels deposited on the cotton-cellulose fibers.

For this purpose, inhibition zone of the gram-negative bacteria *E. coli* and gram-positive bacteria *S. aureus* was determined at temperatures below and above the LCST of poly-NiPAAm, i.e., at 23 and 37 °C (Figure 10). Expectedly, the inhibition zone was not observed for the untreated sample and the CO_M1 and CO_M2 samples, as the bio-barrier formed by the chitosan in the CO_M1 and CO_M2 samples only eliminated the bacteria that came in direct contact with the protonated amino groups of the polymer. By contrast, the CO_M1+S and CO_M2+S samples responded to the temperature changes of the environment, as was indicated by the observed variations in the size of the inhibition zone and the bacteria colonies as the temperature rose from 23 to 37 °C. Namely, when the samples were exposed to temperatures above the LCST of poly-NiPAAm (i.e., 37 °C), the hydrogel particles shrunk and squeezed the savory EO out of their structure. At increased concentration, certain compounds of the savory EO invaded the cell membrane, causing either cell lysis or disruption of adenosine triphosphate synthesis, leading to the cell death [38]. Accordingly, an obvious inhibition zone of approximately 1.5 mm was formed. As seen in Figure 10, there is no distinct difference in the inhibition zone between the CO_M1+S and CO_M2+S samples for both types of bacteria, implying that the increase in the chitosan concentration in the PNCS/CD hydrogels did not affect the proactive release of savory EO.

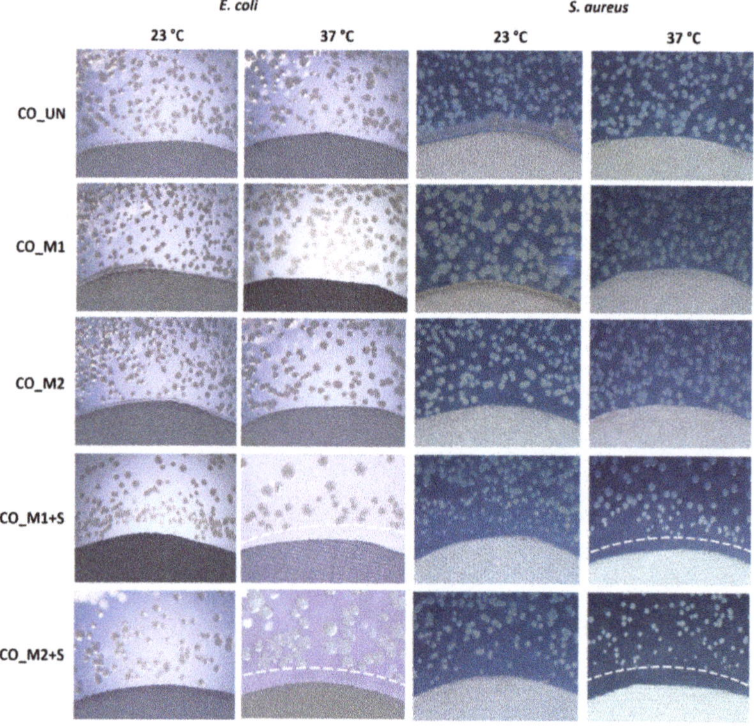

Figure 10. Zone of inhibition formed by the studied cotton-cellulose samples against gram-negative bacteria *Escherichia coli* and gram-positive bacteria *Staphylococcus aureus*, after the incubation at 23 or 37 °C.

On the other hand, when the bacteria were allowed to grow at 23 °C, the CO_M1+S and CO_M2+S samples did not form a distinctive inhibition zone, with bacterial colonies growing in the immediate vicinity of the samples. Nevertheless, from a detailed examination of the results it can be observed that in this case the bacterial colonies clearly changed their morphology and size, indicating that their growth was obstructed. This indicates that the slow leaching of savory EO from the hydrogels in the swollen state must have occurred. Namely, as proposed by Klinger and Landfester [6], the release of the entrapped active substances from the hydrogel matrix does not proceed only through the squeeze-out mechanism, occurring when the hydrogel particles collapse (de-swell) due to the changes in the environment and expel the entrapped active substances from its structure, but may also occur according to the increased diffusivity mechanism when the hydrogel exists in its swollen phase. In this case, the porosity increases due to the enlarged mesh size of the swollen hydrogel particles and the entrapped substances begin to slowly leak from its structure. Therefore, it can be concluded that savory EO was released from the CO_M1+S and CO_M2+S samples according to both of the mechanisms, but the squeeze-out mechanism was clearly dominant.

In continuation of the research, the antimicrobial activity of the samples against *E. coli* and *S. aureus* was analyzed. Based on the number of grown bacteria colonies, the $\log_{10}$CFU/mL was determined (Figure 11). Notably, the higher the $\log_{10}$CFU/mL values, the greater the antibacterial activity of the studied coating against tested bacteria. The results show that the chemical composition of the hydrogels alone had a strong effect on the bacterial reduction, as higher concentration of chitosan in the CO_M2 sample lead to excellent antibacterial activity against both tested bacteria, while the CO_M1 sample showed to be ineffective. In accordance with the results of the zone of inhibition determination, the CO_M1+S and CO_M2+S samples showed increased antibacterial activity against both, *E. coli* and *S. aureus*. In this manner, both CO_M1+S and CO_M2+S samples exhibited total bacterial reduction after 1 h of exposure to the functionalized cotton samples. Comparison of antibacterial activity of these sample shows a strong increase of $\log_{10}$CFU/mL value of the CO_M2+S sample, with antibacterial activity against both tested bacteria greater than 7.0 log. These results prove the achievement of the proposed synergistic antimicrobial action between the chitosan in the PNCS/CD M2 hydrogel and embedded savory EO, as double bio-barrier and controlled release antimicrobial activity was successfully obtained.

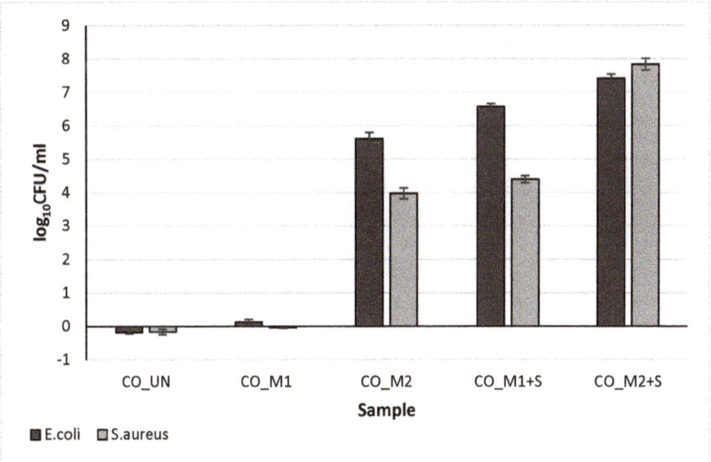

Figure 11. Antimicrobial activity of the studied samples based on the number of grown bacteria colonies right after the exposure to the studied samples and after 1 h exposure with the samples, determined for *E. coli* and *S. aureus*.

4. Conclusions

In this work, "smart," temperature and pH responsive cotton-cellulose fabric with dual antimicrobial activity was tailored. It was shown that chemical composition of the PNCS/CD hydrogels influences the hydrodynamic particle size, with a higher concentration of chitosan reducing the particle sizes and improving the hydrogel particle stability. Accordingly, when present on the surface of the cotton-cellulose fibers, PNCS/CD M2 showed superior temperature and pH responsiveness, the latter derived also due to greater concentration of the protonated amino groups of chitosan. Compared to PNCS hydrogel, the inclusion of β-CD in the PNCS/CD hydrogels structure hindered the temperature related release of moisture, due to greater polymer concentration, which decreased the porosity and increased tortuosity of hydrogel particles. Nevertheless, the presence of β-CD in the PNCS/CD M1 and PNCS/CD M2 hydrogels enabled the embedment of savory EO in its host cavity that was then released from its structure in a controlled manner, i.e., at elevated temperatures. Successful release of savory EO improved the antibacterial activity of the cotton-cellulose samples as a result of the synergism between bio-barrier forming chitosan within the hydrogels and the embedded savory EO. Accordingly, PNCS/CD hydrogel functionalized cotton fabric could easily be combined with various EOs, thus providing pro-active moisture management and antimicrobial activity along with other beneficial effects of EOs, customized according to the market demand. Such high-added value materials are of great importance in the field of medical textiles and related healthcare and hygiene textiles.

Supplementary Materials: The following are available online at http://www.mdpi.com/2079-6412/9/4/242/s1, Table S1: Essential oil chromatography sheet records provided by the producer of the used EO (Florihana Distillerie, Caussols, France), Figure S1: Zone of inhibition of disc impregnated with (a)–DMSO, and discs impregnated with (b)–1%, (c)–2%, (d)–3%, (e)–4% and (f)–5% concentration of savory EO in DMSO.

Author Contributions: Investigation, D.Š.; Analysis, D.Š., M.Š., M.M., E.Š., I.G.I.; Writing—Original Draft Preparation, D.Š.; Writing—Review and Editing, B.T., B.S. and I.J.; Supervision, B.T.

Funding: This research was funded by the SLOVENIAN RESEARCH AGENCY (Programme P2-0213, Infrastructural Centre RIC UL-NTF and a grant for the doctoral student D.Š.).

Conflicts of Interest: The authors declare no conflict of interest.

Abbreviations

APS	Ammonium persulfate
ATR	Attenuated total reflection
CFU	Colony-forming units
C_{MC}	Contribution
CO_M1	Cotton fabric treated with PNCS/CD hydrogel M1 (PN:CS = 7:1)
CO_M1+S	Cotton fabric treated with PNCS/CD hydrogel M1 and savory EO emulsion
CO_M2	Cotton fabric treated with PNCS/CD hydrogel M2 (PN:CS = 4:1)
CO_M2+S	Cotton fabric treated with PNCS/CD hydrogel M2 and savory EO emulsion
CO_UN	Untreated cotton fabric
DLS	Dynamic light scattering
DSC	Differential scanning calorimetry
E. coli	Escherichia coli
EO	Essential oil
FT-IR	Fourier-transform infrared spectroscopy
LCST	Lower critical solution temperature
MBA	*N,N*-methylenebisacrylamide
MC	Moisture content
NiPAAm	*N*-isopropylacrylamide
PDI	Polydispersity index
pK_a	Acid dissociation constant
PN:CS	Poly-NiPAAm to chitosan ratio
PNCS hydrogel	Poly-(*N*-isopropylacrylamide)/chitosan based hydrogel
PNCS/CD hydrogel	Poly-(*N*-isopropylacrylamide)/chitosan/β-cyclodextrine based hydrogel
PNCS/CD M1	Poly-(*N*-isopropylacrylamide)/chitosan/β-cyclodextrine based hydrogel with poly-(*N*-isopropylacrylamide) to chitosan ratio 7:1

PNCS/CD M2	Poly-(*N*-isopropylacrylamide)/chitosan/β-cyclodextrine based hydrogel with poly-(*N*-isopropylacrylamide) to chitosan ratio 4:1
poly-NiPAAm	Poly-(*N*-isopropylacrylamide)
rpm	Round per minute
S. aureus	Staphylococcus aureus
SDS	Sodium dodecyl sulfate
SEM	Scanning electron microscopy
WP	Wet pick-up
WU	Water uptake
β-CD	β-cyclodextrine

References

1. Guerra-Rosas, M.I.; Morales-Castro, J.; Cubero-Marquez, M.A.; Salvia-Trujillo, L.; Martín-Belloso, O. Antimicrobial activity of nanoemulsions containing essential oils and high methoxyl pectin during long-term storage. *Food Control* **2017**, *77*, 131–138. [CrossRef]

2. Morais, D.; Guedes, R.; Lopes, M. Antimicrobial approaches for textiles: From research to market. *Materials* **2016**, *9*, 498. [CrossRef]

3. Islam, S.; Shahid, M.; Mohammad, F. Perspectives for natural product based agents derived from industrial plants in textile applications—A review. *J. Clean. Prod.* **2013**, *57*, 2–18. [CrossRef]

4. Solorzano-Santos, F.; Miranda-Novales, M.G. Essential oils from aromatic herbs as antimicrobial agents. *Curr. Opin. Biotechnol.* **2012**, *23*, 136–141. [CrossRef] [PubMed]

5. Dima, C.; Dima, S. Essential oils in foods: Extraction, stabilization, and toxicity. *Curr. Opin. Food Sci.* **2015**, *5*, 29–35. [CrossRef]

6. Klinger, D.; Landfester, K. Stimuli-responsive microgels for the loading and release of functional compounds: Fundamental concepts and application. *Polymer* **2012**, *53*, 5209–5231. [CrossRef]

7. Radu, C.D.; Parteni, O.; Ochiuz, L. Applications of cyclodextrins in medical textiles—Review. *J. Control. Release* **2016**, *224*, 146–157. [CrossRef] [PubMed]

8. Zhang, Y.; Zhang, H.; Wang, F.; Wang, L.X. Preparation and properties of ginger essential oil β-cyclodextrin/chitosan inclusion complexes. *Coatings* **2019**, *8*, 305. [CrossRef]

9. Bashari, A.; Hemmatinejad, N.; Pourjavadi, A. Smart and fragrant garment via surface modification of cotton fabric with cinnamon oil/stimuli responsive pnipaam/chitosan nano hydrogels. *IEEE Trans. Nanobiosci.* **2017**, *16*, 455–462. [CrossRef] [PubMed]

10. Kettle, M.J.; Dierkes, F.; Schaefer, K.; Moeller, M.; Pich, A. Aqueous nanogels modified with cyclodextrin. *Polymer* **2011**, *52*, 1917–1924. [CrossRef]

11. Yi, P.; Wang, Y.; Zhang, S.; Zhan, Y.; Zhang, Y.; Sun, Z.; Li, Y.; He, P. Stimulative nanogels with enhanced thermosensitivity for therapeutic delivery via β-cyclodextrin-induced formation of inclusion complexes. *Carbohydr. Polym.* **2017**, *166*, 219–227. [CrossRef]

12. Jocić, D. Polymer-based smart coatings for comfort in clothing. *Tekstilec* **2016**, *59*, 107–114. [CrossRef]

13. Feyzioglu, G.C.; Tornuk, F. Development of chitosan nanoparticles loaded with summer savory (*Satureja hortensis* L.) essential oil for antimicrobial and antioxidant delivery applications. *LWT Food Sci. Technol.* **2016**, *70*, 104–110. [CrossRef]

14. Raafat, D.; Von Bargen, K.; Haas, A.; Sahl, H.G. Insights into the mode of action of chitosan as an antibacterial compound. *J. Appl. Environ. Microbiol.* **2008**, *74*, 3764–3773. [CrossRef] [PubMed]

15. Tang, R.; Yu, Z.; Zhang, Y.; Qi, C. Synthesis, characterization, and properties of antibacterial dye based on chitosan. *Cellulose* **2016**, *23*, 1741–1749. [CrossRef]

16. Lee, C.F.; Wen, C.J.; Chiu, W.Y. Synthesis of poly(chitosan-nisopropylacrylamide) complex particles with the method of soapless dispersion polymerization. *J. Polym. Sci. Part A Polym. Chem.* **2003**, *41*, 2053–2063. [CrossRef]

17. Štular, D.; Jerman, I.; Mihelčič, M.; Simončič, B.; Tomšič, B. Antimicrobial activity of essential oils and their controlled release from the smart PLA fabric. *IOP Conf. Ser. Mater. Sci. Eng.* **2018**, *460*, 012011. [CrossRef]

18. Schindler, W.D.; Hauser, P.J. *Chemical Finishing of Textiles*; Woodhead: Cambridge, UK, 2004.

19. *ASTM D629-15 Standard Test Methods for Quantitative Analysis of Textiles*; ASTM International: West Conshohocken, PA, USA, 2015.
20. Kulkarni, A.; Tourrette, A.; Warmoeskerkena, M.M.C.G.; Jocic, D. Microgel-based surface modifying system for stimuli-responsive functional finishing of cotton. *Carbohydr. Polym.* **2010**, *82*, 1306–1314. [CrossRef]
21. *ISO 20645 Textile Fabrics Determination of Antibacterial Activity Agar Diffusion Plate Test*; International Organization of Standards: Geneva, Switzerland, 2004.
22. *EN ISO 20743 Textiles—Determination of Antibacterial Activity of Textile Products*; European Committee for Standardization: Brussels, Belgium, 2013.
23. Bashari, A.; Hemmatinejad, N.; Pourjavadi, A. Surface modification of cotton fabric with dual-responsive PNIPAAm/chitosan nano hydrogel. *Polym. Adv. Technol.* **2013**, *24*, 797–806. [CrossRef]
24. Štular, D.; Jerman, I.; Simončič, B.; Grgić, K.; Tomšič, B. Influence of the structure of a bio-barrier forming agent on the stimuli-response and antimicrobial activity of a "smart" non-cytotoxic cotton fabric. *Cellulose* **2018**, *25*, 6231–6245. [CrossRef]
25. Carrillo, F.; Defays, B.; Colom, X. Surface modification of lyocell fibres by graft copolymerisation of thermo-sensitive poly-N-isopropylacrylamide. *Eur. Polym. J.* **2008**, *44*, 4020–4028. [CrossRef]
26. Draczyński, Z.; Flinčec Grgac, S.; Dekanić, T.; Tarbuk, A.; Boguń, M. Implementation of chitosan into cotton fabric. *Tekstilec* **2017**, *60*, 296–301. [CrossRef]
27. Socrates, G. *Infrared and Raman Characteristic Group Frequencies*; John Wiley & Sons: New York, NY, USA, 2001.
28. Gupta, D.; Haile, A. Multifunctional properties of cotton fabric treated with chitosan and carboxymethyl chitosan. *Carbohydr. Polym.* **2007**, *69*, 164–171. [CrossRef]
29. Lee, S.B.; Ha, D.I.; Cho, S.K.; Kim, S.J.; Lee, Y.M. Temperature/pH-sensitive comb-type graft hydrogels composed of chitosan and poly(N-isopropylacrylamide). *J. Appl. Polym. Sci.* **2004**, *92*, 2612–2620. [CrossRef]
30. Pan, Y.V.; Wesley, R.A.; Luginbuhl, R.; Denton, D.D.; Ratner, B.D. Plasma polymerized N-isopropylacrylamide: Synthesis and characterization of a smart thermally responsive coating. *Biomacromolecules* **2001**, *2*, 32–36. [CrossRef]
31. Sun, G.; Zhang, X.Z.; Chu, C.C. Formulation and characterisation of chitosan based hydrogel having both temperature and pH sensitivity. *J. Mater. Sci. Mater. Med.* **2007**, *18*, 1563–1577. [CrossRef]
32. Crupi, V.; Ficarra, R.; Guardo, M.; Majolino, D.; Stancanelli, R.; Venuti, V. UV–vis and FTIR–ATR spectroscopic techniques to study the inclusion complexes of genistein with β-cyclodextrins. *J. Pharm. Biomed. Anal.* **2007**, *44*, 110–117. [CrossRef] [PubMed]
33. Rachmawati, H.; Edityaningrum, C.A.; Mauludin, R. Molecular inclusion complex of curcumin–β-cyclodextrin nanoparticle to enhance curcumin skin permeability from hydrophilic matrix gel. *Aaps Pharm.* **2013**, *4*, 1303–1312. [CrossRef] [PubMed]
34. Wang, H.D.; Chu, L.Y.; Yu, X.Q.; Xie, R.; Yang, M.; Xu, D.; Zhang, J.; Hu, L. Thermosensitive affinity behavior of poly(N-isopropylacrylamide) hydrogels with β-cyclodextrin moieties. *Ind. Eng. Chem. Res.* **2007**, *46*, 1511–1518. [CrossRef]
35. Salmeri, S.; Lacroix, M. Physicochemical properties of alginate/polycaprolactone-based films containing essential oils. *J. Agric. Food Chem.* **2006**, *54*, 10205–10214. [CrossRef]
36. Turki, A.; El Oudiani, A.; Msahli, S.; Sakli, F. Infrared spectra for alfa fibers treated with thymol. *J. Glycobiol.* **2018**, *7*, 2. [CrossRef]
37. Klemm, D.; Heublein, B.; Fink, H.P.; Bohn, A. Cellulose: Fascinating biopolymer and sustainable raw material. *Angew. Chem. Int. Ed.* **2005**, *44*, 3358–3393. [CrossRef]
38. Rieger, K.A.; Schiffman, J.D. Electrospinning an essential oil: Cinnamaldehyde enhances the antimicrobial efficacy of chitosan/poly(ethylene oxide) nanofibers. *Carbohydr. Polym.* **2014**, *113*, 561–568. [PubMed]

Review

Chitosan Coating Applications in Probiotic Microencapsulation

Lavinia-Florina Călinoiu [1], Bianca Eugenia Ştefănescu [1,2], Ioana Delia Pop [3,*], Leon Muntean [4,*] and Dan Cristian Vodnar [1,*]

[1] Department of Food Science and Technology, Institute of Life Sciences, University of Agricultural Sciences and Veterinary Medicine, Cluj-Napoca, Calea Mănăştur 3-5, 400372 Cluj-Napoca, Romania; lavinia.calinoiu@usamvcluj.ro (L.-F.C.); stefanescu.bianca@umfcluj.ro (B.E.S.)

[2] Department of Pharmaceutical Botany, Iuliu Haţieganu University of Medicine and Pharmacy, 12 I. Creangă Street, 400010 Cluj-Napoca, Romania

[3] Department of Exact Sciences, Horticulture Faculty, University of Agricultural Sciences and Veterinary Medicine, Cluj-Napoca, Calea Mănăştur 3-5, 400372 Cluj-Napoca, Romania

[4] Department of Plant Culture, University of Agricultural Sciences and Veterinary Medicine, Cluj-Napoca, Calea Mănăştur 3-5, 400372 Cluj-Napoca, Romania

* Correspondance: popioana@usamvcluj.ro (I.D.P.); leon.muntean@usamvcluj.ro (L.M.); dan.vodnar@usamvcluj.ro (D.C.V.)

Received: 8 February 2019; Accepted: 13 March 2019; Published: 16 March 2019

Abstract: Nowadays, probiotic bacteria are extensively used as health-related components in novel foods with the aim of added-value for the food industry. Ingested probiotic bacteria must resist gastrointestinal exposure, the food matrix, and storage conditions. The recommended methodology for bacteria protection is microencapsulation technology. A key aspect in the advancement of this technology is the encapsulation system. Chitosan compliments the real potential of coating microencapsulation for applications in the food industry due to its physicochemical properties: positive charges via its amino groups (which makes it the only commercially available water-soluble cationic polymer), short-term biodegradability, non-toxicity and biocompatibility with the human body, and antimicrobial and antifungal actions. Chitosan-coated microcapsules have been reported to have a major positive influence on the survival rates of different probiotic bacteria under in vitro gastrointestinal conditions and in the storage stability of different types of food products; therefore, its utilization opens promising routes in the food industry.

Keywords: microencapsulation; probiotic bacteria; chitosan coating; viability; food applications

---

1. Introduction

Nowadays, there is an increase in functional probiotic food demand, as well as waste-derived bioactive compounds re-utilization [1–8], based on the consciousness of consumers regarding their health potential [9,10]. Considering the IndustryARC report [11] from 2018, the global probiotic market is estimated to experience a compound annual growth rate of 5.6% through 2020.

According to the FAO/WHO, probiotics are characterized as living microorganisms which, when ingested in certain amounts, provide health benefits to the host [12]. Some of these health benefits include antagonistic effects against harmful bacteria in humans and immune effects [13]. Their usage positively influences the growth of targeted microorganisms, eliminates harmful bacteria, and boosts the host's naturally occurring defense actions [14]. In 1993, Ziemer and Gibson [15] were amongst the first researchers to sustain the presence of health-related bacteria in soured milk with an impact on intestinal health. For the past two decades, these bioactive ingredients have been at the forefront of many studies [16–18]. Two of the most common types of microbes extensively used as probiotics are the bacteria belonging to the genera *Bifidobacterium* and *Lactobacillus* [13,19–22]. Considering the above, probiotic-enriched food products should reach the recommended level at the time of consumption, which was agreed upon as being 10^6–10^7 CFU (colony forming units) of viable probiotic bacteria per gram of food [23].

Administered probiotics must resist the harsh gastric conditions [24] and reach the colon in sufficient amounts to be able to sustain colonization, and hence to bring positive benefits to the human body [19,25]. Unfortunately, the free bacteria's inability to survive in high numbers during exposure to the host's gastrointestinal (GI) tract's conditions [26] and/or during exposure to oxygen while a functional food product on a shelf represents the main issues with probiotics [27]. Therefore, their efficiency is highly correlated with their quantity and their viability during storage and product shelf-life [28,29].

Microencapsulation represents the main modern solution for preserving probiotic viability. By definition, microencapsulation represents an incorporation process of probiotic bacteria into a specific material or membrane that has the ability to reduce cell injury or cell loss, derived from environmental factors, with a controlled-release rate under specific conditions [30,31]. Therefore, this technique has been extensively studied during the last decade, since it can maintain the beneficial properties even for sensitive bacteria during storage and absorption [25]. Many studies and reviews have been conducted to investigate and summarize the protective role of this technique [32–35].

Based on the literature available so far, chitosan appears to be one of the most promising coating materials among the most common polymers used for microencapsulation to improve the stability of probiotics [31,36–39]. Moreover, chitosan has a significant protective role against external damages in food products. The antimicrobial ability of chitosan has been observed in numerous studies, of which some resulted in the creation of biodegradable labels, such as the one obtained with chitosan and green tea extract which presents a decontamination effect on the surface of studied fruits and vegetables [40]. Another study even showed its ability to extend the validity of fruit products [41]. Since chitosan is a biopolymer with no or very little sensory influence on food, and considering all the above-mentioned findings, it presents applicability in the food industry [42,43].

The existing literature highlights the high interest in probiotic bacteria microencapsulation, and the importance of coatings for efficient protection and an increased number of probiotics in the GI tract. For the current review, we extracted, evaluated, interpreted, and summarized data related to future trends and implications for applications of chitosan as coating material in probiotic microencapsulation, its performance efficiency on maintaining probiotic viability, protection, and intestinal delivery, as well as its food incorporation aspects.

2. Coatings for Probiotic Microencapsulation

The encapsulation matrix must be food grade and possess suitable physical and chemical properties to deliver protection for the incorporated bacteria [44]. Selection of capsule materials and suitable techniques for tailoring probiotic microcapsules is crucial because it confers the final morphological and functional characteristics of the probiotics [39]. According to Krasaekoopt et al. [45], polymer coatings can significantly increase the chemical and mechanical stability, therefore improving the performance of the microencapsulation materials. Regarding the technologies applied for microencapsulation, emulsion, spray-drying, layer-by-layer (LbL), and extrusion are extensively used and applied at both the laboratory and industrial scales [46–49]. In coating-based encapsulation technology, a major importance is to control the permeability of the coating. Therefore, the LbL approach is a recommended technique since it sustains the permeation of small molecules, while it traps larger molecules. Moreover, the semi-permeable nature of LbL-based coatings can be regulated by the experimental parameters upon assembly [50]. It is important to keep in mind that a combination of these technologies is applied frequently for a higher rate of success.

According to the literature, food-grade coatings like bio-polymers (i.e., alginate, chitosan, pectin, starch, carrageenan, and milk proteins) are the most suitable materials for bacteria microencapsulation due to their high protective rate under certain stress conditions (e.g., gastric pH, bile salts, enzymes) by creating effective physical barriers. Their availability, low-cost, and biocompatibility are major advantages [51–53]. Other compounds, such as proteins and lipids with or without addition of plasticizers and/or surfactants have been proposed and tested as coating materials [33]. The polysaccharide coatings have the ability to prevent oxygen, odor, and oil from entering the capsule (possessing important mechanical characteristics); but, due to their hydrophilic properties, polysaccharides have a big disadvantage, namely moisture permeability [54]. The abovementioned coating materials have been used in combination with alginate-based encapsulation matrices to improve the viability of *Lactobacillus* and *Bifidobacterium* spp. during exposure to acidic conditions. In particular, the alginate–chitosan combination provided efficient protection due to chitosan's strong cationic nature in relation to the anionic alginate [36]. In Table 1 below, a comparison is provided between several coatings for microencapsulation of probiotic bacteria reporting the pros and the cons of each matrix.

Table 1. Type of coatings for microencapsulation of probiotic bacteria: pros and cons.

Coating	Core	Technique	Pros	Cons	Ref.
Chitosan	Alginate, pectin	Extrusion, layer-by-layer (LbL), Emulsion	unique cationic property and high resistance to acidic environment; excellent film-forming abilities; high biocompatibility with living cells and broad antimicrobial activity; tolerance against the deteriorative effects of calcium chelating and anti-gelling agent; dens and strong beads	increases the excretion of sterols and produces a reduction in the digestibility of ileal fats; reported to have inhibitory effects on lactic acid bacteria (LAB) as core material	[31,37,45,55–57]
Alginate	Pectin	Extrusion	simplicity, non-toxicity, biocompatibility and low cost	sensitive in acidic environment; low stability in the presence of chelating agents	[37,45,54,56]
Resistant starch (corn, potato, cassava etc.)	Alginate	Extrusion, emulsion	inexpensive, abundant, biodegradable and easy to use; transparent, odorless, tasteless and colorless; low permeability to oxygen at low-to-intermediate relative humidity; resistant to pancreatic enzymes (amylases), therefore provides good enteric delivery characteristic; is an ideal surface for the adherence of the probiotic cells to the starch granules and this can enhance probiotic delivery	too high viscosity in solution for most of the encapsulation processes	[58–60]
Gelatin	Alginate, pectin	Extrusion	able to form complexes with anionic polymers, such as pectin and alginate	very soluble in aqueous systems	[31,37,61]
Whey protein	Pectin, alginate	Fluidized bed, extrusion	great gelation properties; biocompatible with probiotics; high nutritional value; improvement in the survival of probiotic after exposure to gastric conditions.	difficult to master, longer duration; do not confer additional protection to probiotics when exposed to simulated intestinal conditions	[62–64]
Poly-l-lysine (PLL)	Alginate		food-grade status, active properties and charged behavior	high porosity; does not have a strong capacity to be used as a microcapsule coating for probiotics protection against harsh media	[65–67]
Glucomannan	Sodium alginate		is abundant in nature; is not hydrolyzed by human digestive enzymes	is a non-ionic polymer, so any coating would have to be the result of a non-ionic interaction, such as hydrogen bonding; there is very little work using glucomannan as a coating material for beads	[68,69]
Shellac	Sodium alginate	Fluid bed	natural origin, therefore acceptable as coating material for food supplement products; good resistance to gastric fluid	the low solubility of shellac in the intestinal fluid, especially in the case of enteric coating of hydrophobic substances	[70,71]
Cellulose acetate phthalate (CAP)	Alginate	Emulsion	insoluble in acid media (pH $\leq$ 5) but it is soluble when the pH is $\geq$6 as a result of the presence of phthalate groups resulting in an effective way of delivering large numbers of viable bacterial cells to the colon		[44]
k-carrageenan	Milk, alginate	Extrusion, emulsion	low susceptibility to the organic acids, good efficiency in lactic fermented products (such as yogurt); natural products	dissolves only at high temperatures (60–80 °C) for 2%–5% concentration; irregular shapes and poor mechanical characteristics; the produced gels are not able to withstand stresses	[33,37,72]

The type of coating material has a very significant role in microencapsulation, for example, glucomannan was by far the less effective of all, whereas chitosan coating was reported to provide better protection in simulated gastric conditions than poly-l-lysine (PLL) or alginate coating [73]. The double-layer coating was shown to be significantly better than the single-layer coating. Since each individual coating material possesses some unique, but limited functions, a combination of different encapsulation materials can be more effective.

3. Chitosan-Based Coating Microencapsulation

The microcapsule should be stable and retain its integrity throughout the digestive tract passage until it arrives at its target destination, where the capsule should disintegrate and release its contents [74]. Coating adds an extra protective layer on the microcapsule surface, therefore resulting in improved mechanical strength and a strong barrier function. This process involves the immersion of the hydrogel particles into a solution of coating polymer [35]. The main advantages of chitosan coating are unique cationic character, high biocompatibility, non-toxicity, and biodegradability; therefore, it is quite suitable for use in the food and pharmaceutical industries. Its origin lies in the shell waste of crab, shrimp, and crawfish [25]. The origin source influences the molecular weight of chitosan, which is responsible for its crystallinity, degradation, tensile strength, and moisture content, but can be decreased with processing for increasing the deacetylation [75]. This type of coating is of a major interest in the field of targeted release of probiotics due to its high compatibility with living cells [76]. Chemically speaking, chitosan (Figure 1) is a polysaccharide composed of (1, 4)-linked 2-amino-deoxy-b-d-glucan, a deacetylated derivative of chitin. Chitosan ranges second after cellulose in terms of its availability in nature [77]. The degree of deacetylation of chitin represents the major aspect in chitosan's characterization [78]. For example, when the degree of deacetylation of chitin overcomes 50%, chitosan becomes soluble in aqueous acidic conditions [79]. Moreover, the homogeneous or heterogeneous deacetylation conditions have an important impact on chitosan's microstructure [80], which mainly determines its solubility and applications (i.e., drug or food carriers) [78]. Figure 1 below illustrates the differences between chitin and chitosan, as chemical structures.

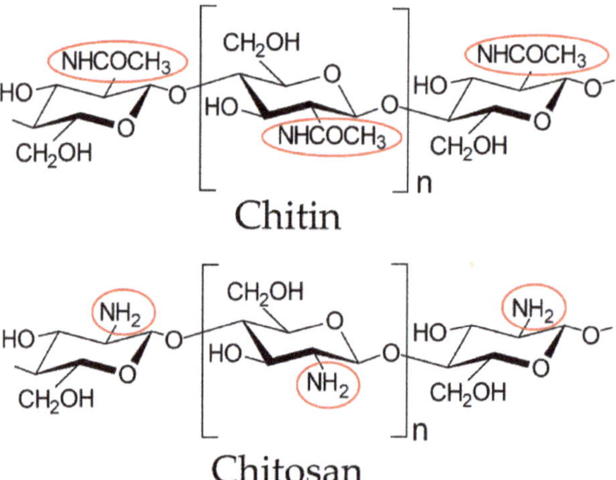

Figure 1. Comparison between the chemical structures of fully acetylated chitin and fully deacetylated chitosan.

It has been reported [73] that coating alginate beads with chitosan develops a complexation of chitosan with alginate resulting in several important properties, such as alginate beads with reduced porosity, reduced leakage of the encapsulated bacteria, and stability at various pH ranges. The negatively charge property of alginate in contact with the positive charge of chitosan develops a semi-permeable membrane, therefore the resulting capsules possess a smoother surface with a reduced permeability to water soluble molecules [73]. However, since the survival of probiotic cells was shown to not be satisfactory and it was reported to have an inhibitory effect against some bacteria (*L. lactis*) [55], chitosan is mostly used as a coating/shell, and not as the capsule itself [81]. In fact, encapsulation of probiotic bacteria with chitosan and alginate coating provides protection in simulated GI conditions, and it is a good way of delivering viable bacterial cells to the colon [37]. Considering the above, chitosan-coated alginate microspheres represents a good alternative for probiotic microorganism oral delivery [82]. The chitosan-specific chemical structure allows important changes at the C-2 position with no difficulties [79]. Based on its aqueous acidic solubility, it allows for many applications in the solution and hydrogel fields, due to its gel-forming abilities. The electrical properties such as the surface potential (ζ-potential) of chitosan-coated alginate microgels or other types of chitosan-based microgels can be evaluated by different methods, e.g., electrophoretic light scattering. For instance, a study [83] published in 2016 evaluated the alginate and chitosan microgels for *B. longum* encapsulation. The particle size of the microbeads was also evaluated using static light scattering, resulting in a higher particle size of the chitosan-coated alginate beads due to the additional coating of alginate or because of some aggregation of the microgels [83].

3.1. Effectiveness of Improving Cell Survival

In order to improve the effectiveness of bacteria survival, researchers have focused on several microencapsulation technologies considering novel combinations of supporting matrices. Several studies conducted on different bacterial strains [84–86] have shown that the use of chitosan-coated microcapsules significantly contributes to the survival of probiotic bacteria during simulated GI conditions. Therefore, experimental studies reported the chitosan-coated alginate microcapsules as the best technology for probiotic bacteria protection (such as *Lactobacillus* and *Bifidobacterium* spp.) against all conditions tested [73,84]. Another study demonstrated that *L. bulgaricus* immobilized by chitosan-coated alginate microencapsulation proved increased storage stability in comparison to free cells [85]. A similar effect was observed in a study conducted by Vodnar and Socaciu [86] on *L. casei* and *L. plantarum*. Moreover, another study highlighted that chitosan coating provided the best protection of probiotic bacteria under simulated GI conditions and their survival increased ($p < 0.05$). A recent study [87] from 2017 showed that pectin–chitosan capsules can protect *L. casei* from the acidic conditions of the stomach and resulted in higher number of viable cells in the intestine [87]. These results are in line with other studies that reported that there was a correlation between the increased concentration of microencapsulating material and the increase in the survival rate of probiotic bacteria under simulated GI conditions [88].

Additionally, it is considered that the probiotic's efficiency and efficacy can be improved by a combination between probiotics and their growth substrate–prebiotics, by a significant colonization of cells in the human gut, since these non-absorbable carbohydrates are a selective energy source for probiotics [86]. This combination was termed as "synbiotics" [89]. Addition of a prebiotic matrix is a promising approach for effective probiotic protection. Therefore, several studies proposing the probiotic–prebiotic chitosan-coated encapsulation system are described below. A simple representation of the concept is illustrated in Figure 2.

In the study by Varankovich et al. [90], the novel pea protein–alginate microcapsules with a chitosan coating were produced by extrusion. These microcapsules were tested for immobilization and survivability of *L. rhamnosus* R0011 and *L. helveticus* R0052 during storage and exposure to in vitro GI conditions. The results indicated the chitosan coating was responsible for an increased cell viability during nine weeks of storage at room temperature, significantly improving the microcapsule performance when compared to non-chitosan coated microcapsules. Under GI conditions, the microcapsule formulation provided high protection for cells, while refrigerated storage had no negative effects on the microcapsule protection performance. In addition, the chitosan coating did not increase the microcapsule size.

Different investigations have demonstrated that selenium-enriched green tea co-encapsulated with probiotic bacteria in chitosan-coated alginate beads offer a compelling approach to expanding the lifespan and viability of probiotic cells in simulated GI juices and refrigerated storage [84,86]. The study by Vodnar and Socaciu [86] on the survival of probiotic bacteria belonging to *L. casei* and *L. plantarum* strains tested during storage at 4 °C demonstrated significantly higher numbers ($p < 0.05$) of survival bacteria encapsulated in chitosan-coated microspheres with selenium-enriched green tea (2 g/100 mL). These results, together with previous findings [84], suggest that immobilization of bacterial strains in chitosan coating improve their viability during refrigeration storage. The chitosan exerts a protective effect on these living microorganisms and the microencapsulation with selenium-enriched green tea was complementary in maintaining the bacteria stability and increased their viability by storage at refrigeration temperature for 30 days. The protective effect of green tea was further demonstrated by sustaining the growth of *Lactobacillus* ssp. and *Bifidobacterium* ssp. during simulated conditions [91].

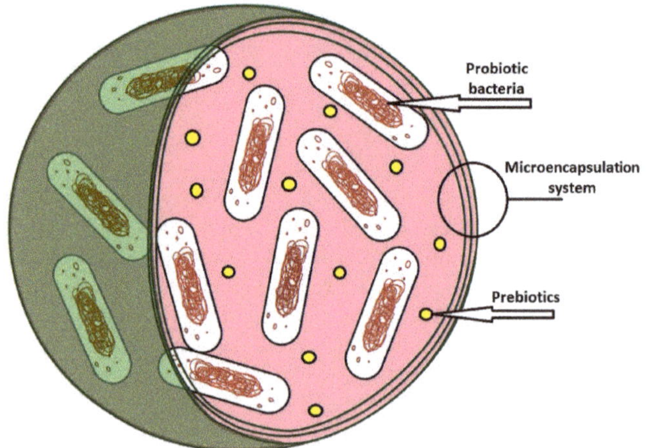

Figure 2. Schematic representation of synbiotics microencapsulation.

Chavarri et al. [82] microencapsulated *L. gasseri* and *B. bifidum* using quercetin as prebiotics and chitosan as coating material in alginate microparticles and reported improved survival during in vitro gastrointestinal conditions. Other studies reported resistant starch as prebiotics and chitosan as coating material for encapsulation of different probiotic bacteria and found increased viability up to 6 months at room temperature [65].

According to de Araújo Etchepare et al. [58], the use of the prebiotic Hi–maize (1%) and chitosan (0.4%) in alginate beads by extrusion technique significantly improved the viability of the microencapsulated bacteria *L. acidophilus* in both the GI and storage conditions of moist and freeze-dried microcapsules.

In a study by Jantarathin et al. [92], *L. acidophilus* TISTR 1338 was separately co-encapsulated with two types of prebiotics, inulin and Jerusalem artichoke, within a chitosan-coated sodium–alginate matrix. After testing the capsules' performance in freeze-drying and high-temperature conditions, the results showed an increase in cells' viability in chitosan double-coated microcapsules, and this increase was maintained after the freeze-dry process. The high-temperature conditions involved the capsules' exposure to 70 °C for 60 min and 90 °C for 5 min, and the findings indicated a 3% prebiotic with 3% alginate and 0.8% chitosan as the most efficient combination for increased viability of microcapsules during heat processing, whereas free cells were destroyed. This novel combination could represent an efficient approach for probiotic bacteria protection during functional food processing that involves heating and freeze-dry processes.

As inulin is one of the most used prebiotics, another study tested the influence of different chain lengths, in co-encapsulation with *L. casei* in chitosan-coated alginate beads. The combination of inulin and chitosan-coating proved to enhance cell viability against gastric and bile salt exposure with 2.7–2.9 log reduction for *L. casei*, where long-chain inulin showed the highest survival rate (2.7 log reduction) [93].

Another efficient approach to improving the viability of probiotic bacteria under GI conditions for targeted release proposes the use of chitosan and enteric polymers in the formulation of microencapsulated beads [94]. For instance, *B. animalis* subsp. lactis was incorporated in alginate,

alginate–chitosan, alginate–chitosan–sreteric, and alginate–chitosan–aryl–eze. The results indicated that the use of chitosan and enteric polymers in the formulation of the beads, especially aryl–eze, improved the survival rate of *B. animalis*, while promoting the controlled intestinal delivery of bifidobacteria.

3.2. Microcapsules Size and Protection Performance

The size of microcapsules is of major importance in probiotic protection. Many recent studies on probiotic encapsulation dealth with particle size reduction due to the negative impact of large particle size on the sensorial and textural characteristics of the product [37]. Heidebach et al. [35] showed that a range of size particles between 0.2 mm and 3 mm can provide strong protection for probiotics during GI exposure. A particle size smaller than 100 μm provided the best sensorial properties. Therefore, considering all these, several solutions have been proposed to eliminate these limitations. For example, a spray-drying technique, very accessible in the food industry, can provide small capsules with average diameters below 100 μm at comparably low costs. Besides, there is a direct relationship between adding a chitosan coating and the microcapsules' diameter [95].

Application of a coating material on the microcapsules' surface is among the proposed solutions for increasing their probiotic performance. The coating materials belong to different type of class compound, and, in some cases, can coincide with the capsules' matrix [56]. By interacting with the capsule surface, the coating will create an extra layer on the microcapsule [35], which can be translated into increased probiotic protection. The coating has the ability to reduce the permeability of the capsule, and implicitly the oxygen exposure of the probiotics, therefore increasing their stability under harsh conditions, such as high temperatures and low pH [35,96]. Other authors have used the coatings for establishing new adhesion properties for the microparticles or to optimize the targeted delivery of the cells [97].

For an increased protection performance under different harsh conditions, multiple combinations of different coating materials and techniques have been applied. For instance, the LbL assembly involves the immersion of microcapsules in polymer solution resulting in the coating, while coacervation implies the formation of a coacervate between the microcapsules' surface and a coating. Regarding the coating development, a major aspect for consideration is the control of the layer's thickness, which, according to previous studies, have no influence on the increase of the capsules' size. Cook et al. [98] demonstrated that the thickness of a chitosan-coated alginate microcapsule is directly correlated with the immersion time, with a minimal value of 8 μm after 1 min, and a value of 24 μm after 2400 min, on microcapsules with a diameter of 1 mm.

In a recent review article by Ramos et al. [54], an in-depth comparative analysis of protection performance for different coatings was investigated. The conclusion suggested that the coatings with better protection performance considering hard digestion conditions were chitosan, alginate, poly-L-lysine (PLL), and whey protein; however, among all, chitosan showed the best efficiency due to its ability to resist and protect the probiotic viable cell during in vitro digestion. In addition, the authors' conclusions underlined the idea that more coatings do not always imply better protection when compared to mono-coated microcapsules.

Chitosan demonstrated to be the most satisfying material to protect microencapsulated probiotics, having efficient results in a variety of alginate microcapsules (performed by different techniques and with different types of alginate), probiotics strains, and exposure conditions. The improved capsule stability and efficient protection was due to the strong ionic interactions between alginate (anionic group) and chitosan (cationic group). Figure 3a illustrates more details regarding this process, where initial microcapsules produced by an anionic encapsulation material (e.g., alginate) was consecutively coated by a cationic material (e.g., chitosan) and after that by another anionic material. The electrostatic forces involved, due to the polyelectrolyte properties of the biopolymers, will contribute to the layer formation that will coat the probiotic-loaded microcapsule [97]. Their ionic interaction representation is illustrated in Figure 3b.

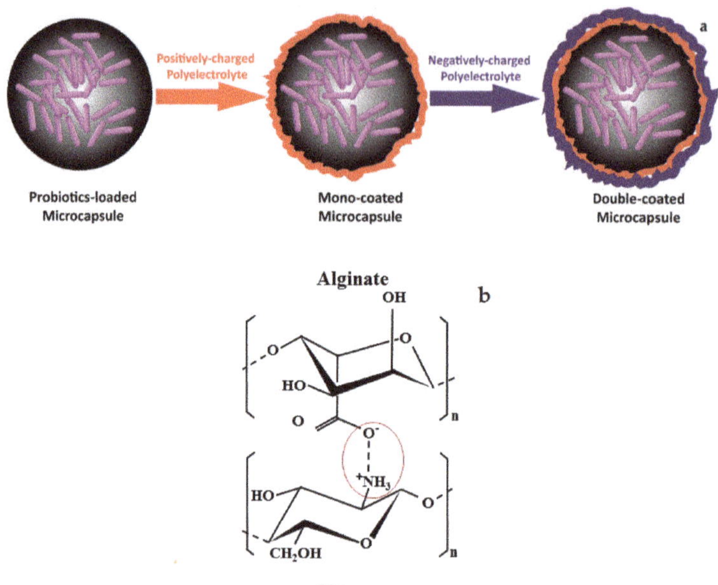

Figure 3. The layer-by-layer (LbL) technique scheme on probiotic microcapsules via coatings (**a**); ionic interaction between alginate and chitosan (**b**).

3.3. Chitosan Application According to the Technology and Bio-Based Matrices Used for the Microencapsulation

Different technologies and core materials can be used to develop probiotic encapsulation with chitosan coatings resulting in microcapsules with different characteristics in terms of range in sizes of particles and of types of capsule, as well as protection efficiency. Novel chitosan bio-based matrices were developed with the aim of an increased bacteria protection making use of the most efficient microencapsulation techniques. In most of the studies, chitosan coating proved to have a high efficiency in different probiotic strain protection against harsh conditions, maintaining a proper concentration of viable cells for intestinal delivery. Regarding technologies, spray-dried particles coated with chitosan are recommended as significantly effective capsules in delivering viable bacterial cells to the colon and stable particles during refrigerated storage [99]. Although, one of the most studied encapsulation technologies regarding chitosan coating is extrusion LbL.

In a study by Singht et al. [100], novel bio-based matrices of carboxymethyl cellulose–chitosan (CMC-Cht) were used for the encapsulation of the probiotic bacteria *L. rhamnosus* GG via a nozzle-spray method. The hybrid micro- and macroparticles results confirmed their potential for encapsulation and delivery, being the first successful encapsulation of *L. rhamnosus* GG in CMC-Cht particles with an acceptable survival rate. Li et al. [101] encapsulated *L. casei* ATCC 393 with alginate, chitosan, and carboxymethyl chitosan matrices by an extrusion method, and the system increased the cells' viability up to 10^8 cfu/g in a dry state after 4 weeks of storage at 4 °C. After exposure to GI conditions, the encapsulated bacteria maintained its probiotic effect, indicating that alginate–chitosan–carboxymethyl chitosan microcapsules could efficiently protect *L. casei* against harsh conditions and may represent a novel route for delivery of probiotic cultures as a functional food.

In a study by Zou et al. [66], the encapsulation of *B. bifidum* F-35 in alginate microspheres developed by emulsification/internal gelation technique was reinforced by addition of pectin/starch or coating with chitosan/PLL to enhance protection for probiotic bacteria. By comparison, the chitosan-coated alginate microspheres showed the highest protection for microencapsulated bacteria under in vitro GI conditions and during 1 month of storage at 4 °C, being an efficient approach for bifidobacteria intestinal colonization.

Cook et al. [102] investigated the LbL coating of alginate matrices with chitosan–alginate for encapsulation of *B. breve* with the aim of improving bacteria survival under low-pH conditions, and implicitly, intestinal delivery. The experimental study proved that multilayer-coated alginate matrices increased cells' viability during exposure to in vitro gastric conditions, precisely from <3 log(CFU)/mL, reported in free cells, up to a maximum of 8.84 ± 0.17 log(CFU)/mL in the 3-layer coated matrix, while also providing a targeted gradual intestinal release over 240 min. There are also other studies reporting that chitosan-coated alginate microparticles for probiotic encapsulations allowed better viability [65,103]. Chitosan and alginate have been tested many times for coating abilities in microencapsulation and protection of different probiotics (such as *B. bifidum*, *B. breve*, and *L. gasseri*) [82,98,102,104]. Chitosan and alginate possess high-charge densities, being able to increase the capsule's residence in targeted areas of release. Therefore, they provide probiotic intestinal delivery [98].

Fareez et al. [105] successfully implemented the microencapsulation of *L. plantarum* LAB12 in chitosan–alginate–xanthan gum–β–cyclodextrin (Alg–XG–β–CD–Ch) beads considering a survival rate of 95% at pH 1.8 with facilitated release at pH 6.8. Moreover, the microcapsules maintained the cells' viability >7 log CFU/g during 4-week storage at 4 °C and had reduced viable cell loss at 75 °C and 90 °C. Considering this, the Alg–XG–β–CD–Ch approach may be suitable for application as heat- and pH-stable polymeric beads that incorporate lactobacilli species as efficient transport vehicles crossing gastric conditions for final intestinal colonization, as heat resistant coating up to 90 °C is a significant property in product manufacturing. Therefore, the Alg–XG–β–CD–Ch applications for probiotics are wide and target the health, food, and agro-industries. In 2015, the same author, Fareez et al. [106], demonstrated that incorporation of the same probiotic bacteria into chitosan-coated alginate–xanthan gum (Alg–XG) beads was a feasible physicochemical driven approach for delivering new functional food ingredients [106].

Falco et al. [107], using the LbL technique, developed a chitosan and sulfated β-glucan encapsulation matrix for *L. acidophilus* considering their prebiotic property for further novel applications, such as carriers for probiotics and sensitive nutraceuticals. Compared to uncoated cells, the viability of cells with four layers of chitosan and sulfated β-glucan decreased only by 2 log CFU/mL. Under in vitro GI exposure, the protection of the coatings was partially degraded, but resisted under acidic gastric conditions. The Hi–maize (1.0% w/v) prebiotic addition to microcapsules containing *Lactobacillus* spp. coated with chitosan considerably improved ($p < 0.05$) the viability of cells after GI exposure, and in stored yogurt, in comparison with alginate-based microcapsules [65].

According to Bepeyeva et al. [87], encapsulation of *L. casei* into calcium–pectinate–chitosan beads provided protection of cells under GI exposure. The beads were prepared by extrusion of amidated pectin into calcium chloride with additional chitosan coating, resulting in high levels of viable bacteria with intestinal delivery application. According to Kanmani et al. [108], the encapsulation of LAB *Enterococcus faecium* MC13 into chitosan-coated alginate microcapsules demonstrated an improved delivery of viable cells and good resistance to harsh gastro-intestinal conditions. Trabelsi et al. [109] reported that encapsulated *L. plantarum* TN8 on alginate coated with chitosan during 8 weeks of storage at 4 °C was effective in maintaining the stability of the probiotic bacteria.

The experimental results of Zaeim et al. [110] proposed wet-electrospraying as a successful and novel technique for encapsulation of probiotic bacteria (*L. plantarum*) inside Ca–alginate/chitosan hydrogel microcapsules by single- and double-stage procedure with an encapsulation yield of almost 98%. The cells' viability increased with 1 log cycle compared to the free cells under simulated GI conditions, while the outer layer of chitosan, which was deposited on Ca–alginate microcapsules by double-stage procedure, more efficiently protected bacteria at low pH environments.

Table 2 below illustrates the survival rate of different probiotic bacteria in chitosan-coated microcapsules prepared by extrusion-LbL technology.

Table 2. Technology-matrix chitosan-coated encapsulation and its applications.

Microencapsulation Technique	Encapsulation Material	Chitosan-Coating	Probiotic Bacteria	Capsule Size (µm)	Application	Survivability (G: gastric; I: intestinal; colony)	Ref.
Extrusion; layer-by-layer (LbL)	Alginate (2%)+ 0.05 M CaCl2	Chitosan (0.4%)	*Bifidobacterium breve* NCIMB 8807	n.a.	In vitro GI exposure	(log colony forming units (CFU)/mL) (G; I) 7.3; 6.8	[108]
Extrusion; LbL	Alginate (2%)+ 0.5 M CaCl2 + galactooligosaccharides and inulin	Chitosan (0.4%)	*Lactobacillus acidophilus* 5 and *Lactobacillus casei* 01	1830–1850	In vitro GI exposure Refrigerated storage for 4 weeks in yogurt and juice	(log CFU/mL) 2.7 and 2.3 > 10^7 CFU/g^{-1}	[111]
Extrusion; LbL	Alginate (1.8%) + 0.1 M CaCl2 + Hi–maize concentration of up to 1.0% (w/v)	Chitosan Poly-l-lysine (PLL) Alginate	*L. acidophilus* CSCC 2400 or CSCC 2409	500	In vitro GI exposure	(log CFU, app) Chitosan: 9.1 PLL: 7.3 Alginate: 6	[65]
Extrusion; LbL	Alginate (2%) + 0.5 M CaCl2	Chitosan (0.7%)	*Lactobacillus reuteri* DSM 17938	110 ± 5	8 days storage in different solutions at 4 and 20 °C In vitro GI exposure Osmotic stress conditions	log CFU/mL) (G; I) 9.15; 9.3	[112]

3.4. Food Applications of Probiotic Microencapsulated in Chitosan-Based Coatings

The most important food applications of chitosan include the encapsulating material for probiotic stability in the production of functional food products [44], formation of biodegradable films, enzymes binding, conservation of foods from microbial deterioration, nutritional supplements, and other applications (additives, emulsifier agents, etc.) [113]. Belonging mainly to lactic acid bacteria (LAB), probiotics are widely used in the production of fermented dairy foods such as yoghurt, cheese, korut, and kefir, being the richest sources of probiotic foods available on the market [39], but in recent years, the focus of using probiotic microencapsulation techniques have moved to fruit juices [114], cereal-based products, chocolate products [115], and cookies—this being a real challenge considering the product matrix [19]. Furthermore, a screening of dairy products, beverages, and other products developed with incorporation of probiotics microencapsulated in chitosan-based coating is presented.

3.4.1. Dairy Products

The microcapsules developed by distinctive technologies with an extra coating represent a technological step recommended to increase protection of the bioactive compounds from external damage factors such as acidity, oxygen, and gastric conditions [25] while incorporated in dairy products. Since the incorporation of microcapsules in yogurt products do not alter the sensory quality [116], chitosan is the perfect candidate for the role of coating material, due to its non-impairing adverse sensory properties to food [117]. In the beginning of the 20th century, the challenge of using chitosan to incorporate LAB was addressed [118], and since then, different bacterial strains were taken under investigation.

One report concluded that *L. delbrueckii* subsp. *bulgaricus* immobilized by chitosan-coated alginate maintained cell stability for 4 weeks of storage at 4 °C and 22 °C in skim milk [119]. Studies performed on strains belonging to *L. bulgaricus*, *L. gasseri*, and *B. bifidum* [82,85] loaded in chitosan-coated alginate microspheres showed higher storage stability than free cell cultures. Moreover, in a previous study [120], a comparison was made between the survival rate of bifidobacteria encapsulated in alginate beads containing chitosan and that of the bacteria immobilized only in alginate beads. The results obtained showed that chitosan-based capsules provided higher protection for probiotic cells than alginate matrix in yogurt products and under simulated GI exposure [120]. The study by Urbanska et al. [121] demonstrated the effectiveness of chitosan-coated alginate microcapsules for delivery of probiotic *L. acidophilus* live cells in yogurt. Moreover, the results reported a structural integrity of microcapsules after 76 h of mechanical agitation in culture broth media and after 24 h in in vitro GI conditions. Krasaekoopt et al. [45] microencapsulated *L. acidophilus* 547, *B. bifidum* ATCC 1994, and *L. casei* 01 in chitosan-coated alginate beads with incorporation in yoghurt from UHT and conventionally treated milk for investigating their survival during storage at 4 °C for 4 weeks. The survival of encapsulated probiotic bacteria was higher than free cells, while the probiotic effect was maintained, the viable cells' number being above the recommended therapeutic level during storage, except for *B. bifidum*.

In formulated yoghurt products, the viability of probiotics was improved by applying sodium alginate beads, which were processed with chitosan as an effective microencapsulation to maintain stability under storage at refrigeration temperature. A four times higher viability in yoghurt-applied capsules compared to cells in a saline suspension was observed [122]. This reinforces the fact that microencapsulation with chitosan coating represents an important alternative. Moreover, it is very effective in providing the colon with higher numbers of viable bacterial cells and keeping their survival in dairy products under refrigeration conditions.

Obradović et al. [123] investigated the protection of chitosan coating on cell viability of microencapsulated probiotic starter culture (containing *S. salivarius* ssp. thermophilus, *L. delbrueckii* ssp. bulgaricus, *L. acidophilus*, and *B. bifidum*) in fermented whey beverages against fermentation process conditions and product storage. The chitosan coating's influence on the mechanical stability of core encapsulation material was also assessed. Sodium–alginate beads were made using the extrusion technique. The results revealed an increased cell viability with chitosan coatings, as well as improved elastic and strength properties of beads during food storage.

The combination of two lactobacilli (*L. acidophilus* and/or *L. reuteri*) were successfully microencapsulated in alginate and alginate–chitosan beads for addition to milk and blackberry jam set-style yogurt [124]. After storage at 5 °C for 30 days, followed by simulated GI conditions exposure, the results indicated that alginate–chitosan encapsulation provided better protection than alginate alone, and increased bacteria survival during storage, with cell counts higher than $\geq 10^7$ CFU/g, while after GI simulation, the alginate–chitosan system prevented lactobacilli loss and had favorable intestinal releases. The presence of capsules in blackberry jam set-style yogurt had no sensorial influence, while it did in milk. These two types of dairy products can promote microcapsule stability and lactobacilli viability.

The new encapsulation system xanthan–chitosan and xanthan–chitosan–xanthan, where chitosan was applied as coating, improved storage stability of *B. bifidum* BB01 in yogurt during 21 days at both 4 and 25 °C, providing high probiotic survival during GI tract conditions [125].

3.4.2. Beverages

One study [126] looked at the effect of multi-layer coating of alginate beads on the viability of immobilized *L. plantarum* under in vitro gastric conditions and during storage in pomegranate juice (a highly acidic juice) at 4 °C. The examined beads were either uncoated, single, or double coated in chitosan. The results obtained showed an improvement in the cells' survival rate in the case of chitosan-coated beads, under simulated gastric solution (pH 1.5) by 0.5–2 logs compared to the control (uncoated beads). The strong protection of chitosan may be the result of electrostatic interactions between chitosan and alginate beads. This was the first study that researched this double-coated process for immobilization of probiotics with the aim of increasing their survival and resistance, proving to be better than the single-coated process [126]. Moreover, in a later study, the same authors [68] confirmed that the use of double-chitosan-coated alginate beads yielded a cell concentration of 10^7 CFU/mL and 10^5 CFU/mL for *L. plantarum* and *B. longum*, respectively, after 6 weeks of storage in pomegranate juice and cranberry juice. Therefore, the chitosan coating offered a significant additional protection to that of the encapsulation matrix on the bacteria during storage of microcapsules inside the juice products. This supports the statement that more than one chitosan-layer coating is a promising approach to be used for improving the survival of probiotic cells in strong acidic food matrices [68].

García-Ceja et al. [124] developed a probiotic peach nectar by addition of microencapsulated *L. acidophilus* and *L. reuteri* in an alginate–chitosan system for efficient protection. The results revealed that alginate–chitosan beads protected lactobacilli viability in acidic peach nectar, thus, representing a strong alternative for functional beverage products considering the combination of two lactobacilli, therefore providing more health benefits to consumers.

As described in all the abovementioned publications, the survival of probiotic bacteria in alginate beads containing chitosan was better than in alginate beads alone; therefore, this indicates that this may be used for enhancing the survival of strains. Moreover, consumer health issues and environmental consciousness play important roles in the design of next generation encapsulation matrices and technology, and since chitosan is biocompatible, non-toxic, and biodegradable, further research on the usage of chitosan as a coating material for probiotics will benefit the development of novel functional food products.

3.4.3. Other Food Products

Malmo et al. [116] developed a probiotic chocolate soufflé with *L. reuteri* DSM 17938 microencapsulation via a chitosan-coated alginate system, incorporating it into the dough matrix prior to baking at 180 °C for 10 min (80 °C in the core of product). The authors reported a survival percentage of 10% of the probiotic population after baking and only 1% for free cells. Moreover, the study showed a significant resistance of microencapsulated bacteria when exposed to high temperatures in real food testing compared to the in vitro conditions, indicating a possible extra-protective layer of the food matrices on probiotic cells.

Microencapsulated *L. acidophilus* LA-5 was successfully incorporated in probiotic jelly dessert by Talebzadeh and Sharifan [127]. When compared to free bacteria and alginate beads, the chitosan-coated alginate beads showed increased physical stability, spherical shape, and metabolic activity in GI testing. Moreover, the number of viable coated bacteria maintained above 6 log (10) CFU/g after 42 days of storage and the probiotic jelly provided high-sensory attributes.

The combination of chitosan coating with calcium–alginate and Hi–maize resistant starch microcapsules via emulsion techniques delivered increased viable probiotics: *L. acidophilus* LA-5 and *L. casei* 431 in baked breads [128]. The authors developed synbiotic bread, namely, hamburger buns and white pan breads by inulin addition. Results showed that this microencapsulation system can be used to develop probiotic bakery products with enhanced cell viability against high-thermal conditions with no negative impact on texture or taste, considering that hamburger buns had a higher probiotic survival rate and *L. casei* 431 was more resistant to high temperature than *L. acidophilus* LA-5.

The most recent study by de Farias et al. [129] used a calcium alginate–chitosan microencapsulation system via extrusion method to incorporate *L. rhamnosus* ASCC 290 and *L. casei* ATCC in yellow mombin ice cream. The authors compared the behavior and viability of free and encapsulated cells inside the food matrix against storage at low-temperature condition (−18 °C for 150 days) and GI exposure. Results revealed that free *L. casei* (−1.64 log) had a higher resistance to freezing than free *L. rhamnosus* (−1.92 log), while encapsulated *L. rhamnosus* and *L. casei* presented protection efficiencies of 73.8% and 79.5%, respectively. In the GI simulation, 86.2% *L. rhamnosus* (−0.83 log) and 84% *L. casei* (−1.3 log) were protected by the alginate–chitosan capsules. Therefore, for preparing probiotic yellow mombin ice cream, the encapsulation process is not advantageous for all probiotic bacteria, namely, *L. rhamnosus*, whose survival rate was higher in free form than in microencapsulation, but advantageous for *L. casei*.

The hydrocolloids used in probiotic microencapsulation is a widely-used method for enhancing survival in ice cream during frozen storage. The study by Zanjani et al. [130] indicates that the microencapsulation of probiotics via calcium alginate, wheat, rice, and high-amylose corn (hylon VII) starches coated by chitosan and PLL enhanced probiotic bacteria survival, namely, *L. casei* ATCC 39392 and *B. adolescentis* ATCC 15703, in ice cream after storage at −30 °C for 100 days. Chitosan and PLL coatings significantly increased cell viability during the storage of ice cream, as well as the size of microcapsules. This is due to the integrated microcapsule structure provided by hylon starch. Moreover, sensory evaluation of probiotic ice cream indicated no significant effect on organoleptic properties during the storage period at −30°.

4. Conclusions and Future Perspectives

There is a constant concern that free bacteria might not survive in sufficient numbers during their passage through the GI tract in order to exert its probiotic effect. The physical protection of probiotics by microencapsulation with chitosan-coated alginate beads is an efficient approach to improve the probiotics' survival during GI passage and to achieve a controlled delivery in the intestine. Moreover, multi-stage coating was shown to further increase bacterial survival in acidic food products.

Since the incorporation of probiotics into food matrices is among the challenging areas of research in food technology, and probiotics are quite sensitive to environmental conditions, such as oxygen, light or temperature, and food matrix interactions, the protection of cells is of major importance for the next generation of probiotic foods. Another major challenge is to improve the viability of probiotics during the manufacturing processes, particularly heat processing while considering the perspective of producing thermoresistant probiotic microorganisms as new solutions needed in future research. Therefore, discovering new strains of probiotic bacteria that are heat resistant, either naturally or which have been genetically modified, and creating a microencapsulation system that acts just as "insulation material" are among the most feasible routes. For developing novel encapsulation systems, there is a need for understanding the thermal conductivity properties of most efficient food-grade biopolymers and lipids that are used as encapsulating core materials and coatings, individually and in combination.

Nevertheless, microencapsulation represents the best alternative since it offers a wide range of food application. In a wider sense, encapsulation may be used for plenty of applications within the food industry, such as: production of novel food products, extending the shelf life of functional products, protecting compounds against nutritional loss, controlling the oxidative reaction during storage, providing sustained or controlled release in the gastro-intestinal environment, maintaining the sensory attributes of probiotic-based food products, formation of biodegradable films, and edible labels.

Due to the abundant amino groups, chitosan provides many positive charges in acidic medium, and represents an efficient biopolymer for microencapsulation and delivery systems for the food industry. Moreover, considering its specific physicochemical attributes, biodegradability, and biocompatibility with human tissues, chitosan compliments the real potential of this technology for applications in the food industry. This biopolymer has no negative effects in the amounts used in food because it is natural, non-toxic, and non-allergenic. As the probiotic–prebiotic synergy is well perceived among future trends, insoluble fibers, like β-glucan, are another bio-based natural polysaccharide source less exploited until now, which is available in high quantities in cereal wastes, has biological activities in the human body, and can be fermented by human gut microbiota. β-glucan and chitosan can represent a future delivery system for bioactive molecules and probiotics being a responsive material suitable for targeted release in the intestine.

Dairy products are the main carriers of probiotics and have led the market for many years, but the continuous interest towards improving lifestyle through nutrition led towards the expansion of functional foods variety (beverages, chocolate bars, etc.); therefore, the legislation frame regarding probiotic foods should allow and sustain manufacturers a more effective probiotic food production. New studies must be carried out in order to assess the impact of the chitosan-coated microencapsulated bacteria into a vast range of non-dairy food products, for favoring the needs of particular groups of consumers such as vegetarians, vegans, and lactose-intolerants. Moreover, a deep investigation into the existing material properties for coated capsule production is of major importance for an efficient protection of the probiotic bacteria.

Certainly, the need for in vivo studies evaluating the viability of the incorporated probiotics under GI conditions for establishing the real level of delivered probiotics, and implicitly, the health effects, is a future research direction. Nevertheless, another research trend in this area is to find industrial encapsulation technologies that guarantee the survival of probiotics. In order to achieve these research goals, an integrated approach that combines microencapsulation techniques suitable for the selected food carriers is one of the solutions, as well as consumer behavior assessments toward novel foods considering their future increased demand. Therefore, nowadays, many studies are focusing on reducing the particle size for non-influence on sensorial and textural properties of the product.

Author Contributions: All authors contributed to the preparation of the manuscript, as follows: Conceptualization and design of the work: L.F.C., D.C.V., and I.D.P.; Writing—Manuscript preparation: L.F.C. and B.E.S.; Writing—review and editing: D.C.V., I.D.P., and L.M. All authors read and approved the final manuscript.

Funding: This work was supported by two grants from the Ministry of Research and Innovation, as follows: CNCS–UEFISCDI, project number PN-III-P1-1.1-TE-2016-0661, within PNCDI III; and MCI-UEFISCDI, project number 37 PFE-2018-2020 "Cresterea performantei institutionale prin mecanisme de consolidare si dezvoltare a directiilor de cercetare din cadrul USAMVCN".

Acknowledgments: We kindly thank Vasile Coman for image support and critical peer review; Bernadette E. Teleky, and Martău Adrian for image support.

Conflicts of Interest: The authors declare that they have no competing interests.

References

1. Vodnar, D.C.; Călinoiu, L.F.; Dulf, F.V.; Ştefănescu, B.E.; Crişan, G.; Socaciu, C. Identification of the bioactive compounds and antioxidant, antimutagenic and antimicrobial activities of thermally processed agro-industrial waste. *Food Chem.* **2017**, *231*, 131–140. [CrossRef] [PubMed]

2. Calinoiu, L.-F.; Mitrea, L.; Precup, G.; Bindea, M.; Rusu, B.; Dulf, F.-V.; Stefanescu, B.-E.; Vodnar, D.-C. Characterization of Grape and Apple Peel Wastes' Bioactive Compounds and Their Increased Bioavailability After Exposure to Thermal Process. *Bull. Univ. Agric. Sci. Vet. Med. Cluj-Napoca-Food Sci. Technol.* **2017**, *74*, 80–89. [CrossRef]

3. Szabo, K.; Cătoi, A.-F.; Vodnar, D.C. Bioactive Compounds Extracted from Tomato Processing by-Products as a Source of Valuable Nutrients. *Plant Foods Hum. Nutr.* **2018**. [CrossRef] [PubMed]

4. Călinoiu, L.F.; Vodnar, D.C. Whole Grains and Phenolic Acids: A Review on Bioactivity, Functionality, Health Benefits and Bioavailability. *Nutrients* **2018**, *10*, 1615. [CrossRef] [PubMed]

5. Călinoiu, L.F.; Mitrea, L.; Precup, G.; Bindea, M.; Rusu, B.; Szabo, K.; Dulf, F.V.; Ştefănescu, B.E.; Vodnar, D.C. Sustainable use of agro-industrial wastes for feeding 10 billion people by 2050. In *Professionals in Food Chains*; Wageningen Academic Publishers: Wageningen, The Netherlands, 2018; pp. 482–486. ISBN 978-90-8686-321-1.

6. Mitrea, L.; Trif, M.; Catoi, A.-F.; Vodnar, D.-C. Utilization of biodiesel derived-glycerol for 1,3-PD and citric acid production. *Microb. Cell Factories* **2017**, *16*, 190. [CrossRef] [PubMed]

7. Dulf, F.V.; Unguresan, M.L.; Vodnar, D.C.; Socaciu, C. Free and Esterified Sterol Distribution in Four Romanian Vegetable Oil. *Not. Bot. Horti Agrobot. Cluj-Napoca* **2010**, *38*, 91–97.

8. Coldea, T.E.R.; Socaciu, C.; Parv, M.; Vodnar, D. Gas-Chromatographic Analysis of Major Volatile Compounds Found in Traditional Fruit Brandies from Transylvania, Romania. *Not. Bot. Horti Agrobot. Cluj-Napoca* **2011**, *39*, 109–116. [CrossRef]

9. Andreicut, A.-D.; Parvu, A.E.; Mot, A.C.; Parvu, M.; Fodor, E.F.; Catoi, A.F.; Feldrihan, V.; Cecan, M.; Irimie, A. Phytochemical Analysis of Anti-Inflammatory and Antioxidant Effects of Mahonia aquifolium Flower and Fruit Extracts. *Oxid. Med. Cell. Longev.* **2018**, *2879793*. [CrossRef]

10. Andreicu, A.-D.; Pârvu, A.E.; Pârvu, M.; Fischer-Fodor, E.; Feldrihan, V.; Florinela, A.; Cecan, M.; Irimie, A. Anti-Inflammatory and Antioxidant Effects of Mahonia Aquifolium Leaves and Bark Extracts. *Farmacia* **2018**, *66*, 49–58.

11. Probiotics Market Research Report: Market size, Industry outlook, Market Forecast, Demand Analysis, Market Share, Market Report 2018–2023. Available online: https://industryarc.com/Report/7492/probiotics-market-analysis.html?gclid=EAIaIQobChMIm--8urnb4AIVyOF3Ch1E7AwNEAAYAiAAEgJlh_D_BwE (accessed on 27 February 2019).

12. Food and Agriculture Organization of the United Nations; World Health Organization. *Probiotics in Food: Health and Nutritional Properties and Guidelines for Evaluation*; Food and Agriculture Organization of the United Nations, World Health Organization: Rome, Italy, 2006; ISBN 978-92-5-105513-7.

13. Solanki, H.K.; Pawar, D.D.; Shah, D.A.; Prajapati, V.D.; Jani, G.K.; Mulla, A.M.; Thakar, P.M. Development of microencapsulation delivery system for long-term preservation of probiotics as biotherapeutics agent. *BioMed Res. Int.* **2013**, *2013*. [CrossRef] [PubMed]

14. Dunne, C. Adaptation of bacteria to the intestinal niche: probiotics and gut disorder. *Inflamm. Bowel Dis.* **2001**, *7*, 136–145. [CrossRef]

15. Ziemer, C.J.; Gibson, G.R. An overview of probiotics, prebiotics and synbiotics in the functional food concept: Perspectives and future strategies. *Proc. Int. Dairy J.* **1998**, *8*, 473–479. [CrossRef]

16. Vodnar, D.C.; Venus, J.; Schneider, R.; Socaciu, C. Lactic Acid Production by Lactobacillus paracasei 168 in Discontinuous Fermentation Using Lucerne Green Juice as Nutrient Substitute. *Chem. Eng. Technol.* **2010**, *33*, 468–474. [CrossRef]

17. Pop, O.L.; Brandau, T.; Schwinn, J.; Vodnar, D.C.; Socaciu, C. The influence of different polymers on viability of Bifidobacterium lactis 300b during encapsulation, freeze-drying and storage. *J. Food Sci. Technol.-Mysore* **2015**, *52*, 4146–4155. [CrossRef]

18. Rotar, A.M.; Vodnar, D.C.; Bunghez, F.; Catunescu, G.M.; Pop, C.R.; Jimborean, M.; Semeniuc, C.A. Effect of Goji Berries and Honey on Lactic Acid Bacteria Viability and Shelf Life Stability of Yoghurt. *Not. Bot. Horti Agrobot. Cluj-Napoca* **2015**, *43*, 196–203. [CrossRef]

19. Rokka, S.; Rantamäki, P. Protecting probiotic bacteria by microencapsulation: Challenges for industrial applications. *Eur. Food Res. Technol.* **2010**, *231*, 1–12. [CrossRef]

20. Capozzi, V.; Arena, M.P.; Crisetti, E.; Spano, G.; Fiocco, D. The hsp 16 Gene of the Probiotic Lactobacillus acidophilus Is Differently Regulated by Salt, High Temperature and Acidic Stresses, as Revealed by Reverse Transcription Quantitative PCR (qRT-PCR) Analysis. *Int. J. Mol. Sci.* **2011**, *12*, 5390–5405. [CrossRef]

21. Taranu, I.; Marin, D.E.; Braicu, C.; Pistol, G.C.; Sorescu, I.; Pruteanu, L.L.; Berindan Neagoe, I.; Vodnar, D.C. In Vitro Transcriptome Response to a Mixture of Lactobacilli Strains in Intestinal Porcine Epithelial Cell Line. *Int. J. Mol. Sci.* **2018**, *19*, 1923. [CrossRef] [PubMed]

22. Calinoiu, L.-F.; Vodnar, D.-C.; Precup, G. The Probiotic Bacteria Viability under Different Conditions. *Bull. Univ. Agric. Sci. Vet. Med. Cluj-Napoca-Food Sci. Technol.* **2016**, *73*, 55–60. [CrossRef]

23. Nazzaro, F.; Fratianni, F.; Coppola, R.; Sada, A.; Orlando, P. Fermentative ability of alginate-prebiotic encapsulated Lactobacillus acidophilus and survival under simulated gastrointestinal conditions. *J. Funct. Foods* **2009**, *1*, 319–323. [CrossRef]

24. Pop, O.L.; Dulf, F.V.; Cuibus, L.; Castro-Giráldez, M.; Fito, P.J.; Vodnar, D.C.; Coman, C.; Socaciu, C.; Suharoschi, R. Characterization of a Sea Buckthorn Extract and Its Effect on Free and Encapsulated Lactobacillus casei. *Int. J. Mol. Sci.* **2017**, *18*, 2513. [CrossRef] [PubMed]

25. Anal, A.K.; Singh, H. Recent advances in microencapsulation of probiotics for industrial applications and targeted delivery. *Trends Food Sci. Technol.* **2007**, *18*, 240–251. [CrossRef]

26. Dimitrellou, D.; Kandylis, P.; Petrović, T.; Dimitrijević-Branković, S.; Lević, S.; Nedović, V.; Kourkoutas, Y. Survival of spray dried microencapsulated Lactobacillus casei ATCC 393 in simulated gastrointestinal conditions and fermented milk. *LWT Food Sci. Technol.* **2016**, *71*, 169–174. [CrossRef]

27. Arena, M.P.; Caggianiello, G.; Russo, P.; Albenzio, M.; Massa, S.; Fiocco, D.; Capozzi, V.; Spano, G. Functional Starters for Functional Yogurt. *Foods* **2015**, *4*, 15–33. [CrossRef] [PubMed]

28. Li, X.Y.; Chen, X.G.; Cha, D.S.; Park, H.J.; Liu, C.S. Microencapsulation of a probiotic bacteria with alginategelatin and its properties. *J. Microencapsul.* **2009**, *26*, 315–324. [CrossRef] [PubMed]

29. Kim, S.J.; Cho, S.Y.; Kim, S.H.; Song, O.J.; Shin, I.S.; Cha, D.S.; Park, H.J. Effect of microencapsulation on viability and other characteristics in Lactobacillus acidophilus ATCC 43121. *LWT Food Sci. Technol.* **2008**, *41*, 493–500. [CrossRef]

30. Desai, K.G.H.; Jin Park, H. Recent Developments in Microencapsulation of Food Ingredients. *Dry. Technol.* **2005**, *23*, 1361–1394. [CrossRef]

31. Pavli, F.; Tassou, C.; Nychas, G.-J.E.; Chorianopoulos, N. Probiotic Incorporation in Edible Films and Coatings: Bioactive Solution for Functional Foods. *Int. J. Mol. Sci.* **2018**, *19*, 150. [CrossRef]

32. Shori, A.B. Microencapsulation Improved Probiotics Survival During Gastric Transit. *HAYATI J. Biosci.* **2017**, *24*, 1–5. [CrossRef]

33. Martín, M.J.; Lara-Villoslada, F.; Ruiz, M.A.; Morales, M.E. Microencapsulation of bacteria: A review of different technologies and their impact on the probiotic effects. *Innov. Food Sci. Emerg. Technol.* **2015**, *27*, 15–25. [CrossRef]

34. Riaz, Q.U.A.; Masud, T. Recent Trends and Applications of Encapsulating Materials for Probiotic Stability. *Crit. Rev. Food Sci. Nutr.* **2013**, *53*, 231–244. [CrossRef] [PubMed]

35. Heidebach, T.; Först, P.; Kulozik, U. Microencapsulation of Probiotic Cells for Food Applications. *Crit. Rev. Food Sci. Nutr.* **2012**, *52*, 291–311. [CrossRef]

36. Cook, M.T.; Tzortzis, G.; Charalampopoulos, D.; Khutoryanskiy, V.V. Microencapsulation of probiotics for gastrointestinal delivery. *J. Controlled Release* **2012**, *162*, 56–67. [CrossRef]

37. Morales, M.E.; Ruiz, M.A. 16—Microencapsulation of probiotic cells: applications in nutraceutic and food industry. In *Nutraceuticals*; Grumezescu, A.M., Ed.; Nanotechnology in the Agri-Food Industry; Academic Press: Cambridge, MA, USA, 2016; pp. 627–668. ISBN 978-0-12-804305-9.

38. Ravi Kumar, M.N.V. A review of chitin and chitosan applications. *React. Funct. Polym.* **2000**, *46*, 1–27. [CrossRef]

39. De Prisco, A.; Mauriello, G. Probiotication of foods: A focus on microencapsulation tool. *Trends Food Sci. Technol.* **2016**, *48*, 27–39. [CrossRef]

40. Nasui, L.; Vodnar, D.; Socaciu, C. Bioactive Labels for Fresh Fruits and Vegetables. *Bull. Univ. Agric. Sci. Vet. Med. Cluj-Napoca Food Sci. Technol.* **2013**, *70*, 75–83. [CrossRef]

41. Dong, H.; Cheng, L.; Tan, J.; Zheng, K.; Jiang, Y. Effects of chitosan coating on quality and shelf life of peeled litchi fruit. *J. Food Eng.* **2004**, *64*, 355–358. [CrossRef]

42. Ye, M.; Neetoo, H.; Chen, H. Control of Listeria monocytogenes on ham steaks by antimicrobials incorporated into chitosan-coated plastic films. *Food Microbiol.* **2008**, *25*, 260–268. [CrossRef]

43. Vodnar, D.C. Inhibition of Listeria monocytogenes ATCC 19115 on ham steak by tea bioactive compounds incorporated into chitosan-coated plastic films. *Chem. Cent. J.* **2012**, *6*, 74. [CrossRef] [PubMed]

44. Burgain, J.; Gaiani, C.; Linder, M.; Scher, J. Encapsulation of probiotic living cells: From laboratory scale to industrial applications. *J. Food Eng.* **2011**, *104*, 467–483. [CrossRef]

45. Krasaekoopt, W.; Bhandari, B.; Deeth, H.C. Survival of probiotics encapsulated in chitosan-coated alginate beads in yoghurt from UHT- and conventionally treated milk during storage. *LWT Food Sci. Technol.* **2006**, *39*, 177–183. [CrossRef]

46. Anselmo, A.C.; McHugh, K.J.; Webster, J.; Langer, R.; Jaklenec, A. Layer-by-Layer Encapsulation of Probiotics for Delivery to the Microbiome. *Adv. Mater.* **2016**, *28*, 9486–9490. [CrossRef]

47. Arslan-Tontul, S.; Erbas, M. Single and double layered microencapsulation of probiotics by spray drying and spray chilling. *LWT Food Sci. Technol.* **2017**, *81*, 160–169. [CrossRef]

48. Ashwar, B.A.; Gani, A.; Gani, A.; Shah, A.; Masoodi, F.A. Production of RS4 from rice starch and its utilization as an encapsulating agent for targeted delivery of probiotics. *Food Chem.* **2018**, *239*, 287–294. [CrossRef]

49. Huq, T.; Khan, A.; Khan, R.A.; Riedl, B.; Lacroix, M. Encapsulation of Probiotic Bacteria in Biopolymeric System. *Crit. Rev. Food Sci. Nutr.* **2013**, *53*, 909–916. [CrossRef]

50. Schönhoff, M. Layered polyelectrolyte complexes: physics of formation and molecular properties. *J. Phys. Condens. Matter* **2003**, *15*, R1781–R1808. [CrossRef]

51. Valencia-Chamorro, S.A.; Palou, L.; del Río, M.A.; Pérez-Gago, M.B. Antimicrobial Edible Films and Coatings for Fresh and Minimally Processed Fruits and Vegetables: A Review. *Crit. Rev. Food Sci. Nutr.* **2011**, *51*, 872–900. [CrossRef] [PubMed]

52. Cerqueira, M.A.; Bourbon, A.I.; Pinheiro, A.C.; Martins, J.T.; Souza, B.W.S.; Teixeira, J.A.; Vicente, A.A. Galactomannans use in the development of edible films/coatings for food applications. *Trends Food Sci. Technol.* **2011**, *22*, 662–671. [CrossRef]

53. Šuput, D.Z.; Lazić, V.L.; Popović, S.Z.; Hromiš, N.M. Edible films and coatings: Sources, properties and application. *Food Feed Res.* **2015**, *42*, 11–22. [CrossRef]

54. Ramos, P.E.; Cerqueira, M.A.; Teixeira, J.A.; Vicente, A.A. Physiological protection of probiotic microcapsules by coatings. *Crit. Rev. Food Sci. Nutr.* **2018**, *58*, 1864–1877. [CrossRef]

55. Groboillot, A.F.; Champagne, C.P.; Darling, G.D.; Poncelet, D.; Neufeld, R.J. Membrane formation by interfacial cross-linking of chitosan for microencapsulation of Lactococcus lactis. *Biotechnol. Bioeng.* **1993**, *42*, 1157–1163. [CrossRef]

56. Krasaekoopt, W.; Bhandari, B.; Deeth, H. Evaluation of encapsulation techniques of probiotics for yoghurt. *Int. Dairy J.* **2003**, *13*, 3–13. [CrossRef]

57. Raafat, D.; Sahl, H.-G. Chitosan and its antimicrobial potential – a critical literature survey. *Microb. Biotechnol.* **2009**, *2*, 186–201. [CrossRef]

58. de Araújo Etchepare, M.; Raddatz, G.C.; de Moraes Flores, É.M.; Zepka, L.Q.; Jacob-Lopes, E.; Barin, J.S.; Ferreira Grosso, C.R.; de Menezes, C.R. Effect of resistant starch and chitosan on survival of Lactobacillus acidophilus microencapsulated with sodium alginate. *LWT Food Sci. Technol.* **2016**, *65*, 511–517. [CrossRef]

59. Shah, U.; Naqash, F.; Gani, A.; Masoodi, F.A. Art and Science behind Modified Starch Edible Films and Coatings: A Review. *Compr. Rev. Food Sci. Food Saf.* **2016**, *15*, 568–580. [CrossRef]

60. Gharsallaoui, A.; Roudaut, G.; Chambin, O.; Voilley, A.; Saurel, R. Applications of spray-drying in microencapsulation of food ingredients: An overview. *Food Res. Int.* **2007**, *40*, 1107–1121. [CrossRef]

61. Saravanan, M.; Rao, K.P. Pectin–gelatin and alginate–gelatin complex coacervation for controlled drug delivery: Influence of anionic polysaccharides and drugs being encapsulated on physicochemical properties of microcapsules. *Carbohydr. Polym.* **2010**, *80*, 808–816. [CrossRef]

62. Weinbreck, F.; Bodnár, I.; Marco, M.L. Can encapsulation lengthen the shelf-life of probiotic bacteria in dry products? *Int. J. Food Microbiol.* **2010**, *136*, 364–367. [CrossRef]

63. Doherty, S.B.; Gee, V.L.; Ross, R.P.; Stanton, C.; Fitzgerald, G.F.; Brodkorb, A. Development and characterisation of whey protein micro-beads as potential matrices for probiotic protection. *Food Hydrocoll.* **2011**, *25*, 1604–1617. [CrossRef]

64. Gunasekaran, S.; Ko, S.; Xiao, L. Use of whey proteins for encapsulation and controlled delivery applications. *J. Food Eng.* **2007**, *83*, 31–40. [CrossRef]

65. Iyer, C.; Kailasapathy, K. Effect of co-encapsulation of probiotics with prebiotics on increasing the viability of encapsulated bacteria under in vitro acidic and bile salt conditions and in yogurt. *J. Food Sci.* **2005**.

66. Zou, Q.; Zhao, J.; Liu, X.; Tian, F.; Zhang, H.; Zhang, H.; Chen, W. Microencapsulation of Bifidobacterium bifidum F-35 in reinforced alginate microspheres prepared by emulsification/internal gelation. *Int. J. Food Sci. Technol.* **2011**, *46*, 1672–1678. [CrossRef]

67. Cui, J.-H.; Goh, J.-S.; Kim, P.-H.; Choi, S.-H.; Lee, B.-J. Survival and stability of bifidobacteria loaded in alginate poly-l-lysine microparticles. *Int. J. Pharm.* **2000**, *210*, 51–59. [CrossRef]

68. Nualkaekul, S.; Cook, M.T.; Khutoryanskiy, V.V.; Charalampopoulos, D. Influence of encapsulation and coating materials on the survival of Lactobacillus plantarum and Bifidobacterium longum in fruit juices. *Food Res. Int.* **2013**, *53*, 304–311. [CrossRef]

69. Woo, J.-W.; Roh, H.-J.; Park, H.-D.; Ji, C.-I.; Lee, Y.-B.; Kim, S.-B. Sphericity Optimization of Calcium Alginate Gel Beads and the Effects of Processing Conditions on Their Physical Properties. *Food Sci. Biotechnol.* **2007**, *16*, 715–721.

70. Stummer, S.; Salar-Behzadi, S.; Unger, F.M.; Oelzant, S.; Penning, M.; Viernstein, H. Application of shellac for the development of probiotic formulations. *Food Res. Int.* **2010**, *43*, 1312–1320. [CrossRef]

71. Buch, K.; Penning, M.; Wächtersbach, E.; Maskos, M.; Langguth, P. Investigation of various shellac grades: additional analysis for identity. *Drug Dev. Ind. Pharm.* **2009**, *35*, 694–703. [CrossRef]

72. Shi, L.-E.; Li, Z.-H.; Zhang, Z.-L.; Zhang, T.-T.; Yu, W.-M.; Zhou, M.-L.; Tang, Z.-X. Encapsulation of Lactobacillus bulgaricus in carrageenan-locust bean gum coated milk microspheres with double layer structure. *LWT Food Sci. Technol.* **2013**, *54*, 147–151. [CrossRef]

73. Krasaekoopt, W.; Bhandari, B.; Deeth, H. The influence of coating materials on some properties of alginate beads and survivability of microencapsulated probiotic bacteria. *Int. Dairy J.* **2004**, *14*, 737–743. [CrossRef]

74. Vodnar, D.C.; Socaciu, C.; Rotar, A.M.; Stănilă, A. Morphology, FTIR fingerprint and survivability of encapsulated lactic bacteria (Streptococcus thermophilus and Lactobacillus delbrueckii subsp. bulgaricus) in simulated gastric juice and intestinal juice. *Int. J. Food Sci. Technol.* **2010**, *45*, 2345–2351. [CrossRef]

75. Yuan, Y.; Chesnutt, B.M.; Haggard, W.O.; Bumgardner, J.D. Deacetylation of Chitosan: Material Characterization and in vitro Evaluation via Albumin Adsorption and Pre-Osteoblastic Cell Cultures. *Materials* **2011**, *4*, 1399–1416. [CrossRef]

76. de Vos, P.; Faas, M.M.; Spasojevic, M.; Sikkema, J. Encapsulation for preservation of functionality and targeted delivery of bioactive food components. *Int. Dairy J.* **2010**, *20*, 292–302. [CrossRef]

77. Aider, M. LWT Food Science and Technology Chitosan application for active bio-based films production and potential in the food industry: Review. *LWT Food Sci. Technol.* **2010**, *43*, 837–842. [CrossRef]

78. Estevinho, B.N.; Rocha, F.; Santos, L.; Alves, A. Microencapsulation with chitosan by spray drying for industry applications—A review. *Trends Food Sci. Technol.* **2013**, *31*, 138–155. [CrossRef]

79. Rinaudo, M. Chitin and chitosan: Properties and applications. *Prog. Polym. Sci. Oxf.* **2006**, *31*, 603–632. [CrossRef]

80. George, A.; Roberts, F. Structure of chitin and chitosan. In *Chitin Chemistry*; Palgrave: London, UK, 1992; pp. 85–91.

81. Mortazavian, A.; Razavi, S.H.; Ehsani, M.R.; Sohrabvandi, S. Principles and Methods of Microencapsulation of Probiotic Microorganisms. *Iran. J. Biotechnol.* **2007**, *5*, 1–18.

82. Chávarri, M.; Marañón, I.; Ares, R.; Ibáñez, F.C.; Marzo, F.; del Carmen Villarán, M. Microencapsulation of a probiotic and prebiotic in alginate-chitosan capsules improves survival in simulated gastro-intestinal conditions. *Int. J. Food Microbiol.* **2010**, *142*, 185–189.

83. Yeung, T.W.; Üçok, E.F.; Tiani, K.A.; McClements, D.J.; Sela, D.A. Microencapsulation in alginate and chitosan microgels to enhance viability of Bifidobacterium longum for oral delivery. *Front. Microbiol.* **2016**, *7*. [CrossRef]

84. Vodnar, D.C.; Socaciu, C. Green tea increases the survival yield of Bifidobacteria in simulated gastrointestinal environment and during refrigerated conditions. *Chem. Cent. J.* **2012**, *6*, 61. [CrossRef]

85. Koo, S.M.; Cho, Y.H.; Huh, C.S.; Baek, Y.J.; Park, J. Improvement of the stability of Lactobacillus casei YIT 9018 by microencapsulation using alginate and chitosan. *J. Microbiol. Biotechnol.* **2001**, *11*, 376–383.

86. Vodnar, D.C.; Socaciu, C. Selenium enriched green tea increase stability of Lactobacillus casei and Lactobacillus plantarum in chitosan coated alginate microcapsules during exposure to simulated gastrointestinal and refrigerated conditions. *LWT Food Sci. Technol.* **2014**, *57*, 406–411. [CrossRef]

87. Bepeyeva, A.; de Barros, J.M.S.; Albadran, H.; Kakimov, A.K.; Kakimova, Z.K.; Charalampopoulos, D.; Khutoryanskiy, V.V. Encapsulation of *Lactobacillus casei* into Calcium Pectinate-Chitosan Beads for Enteric Delivery. *J. Food Sci.* **2017**, *82*, 2954–2959. [CrossRef]

88. Anekella, K.; Orsat, V. Optimization of microencapsulation of probiotics in raspberry juice by spray drying. *LWT Food Sci. Technol.* **2013**, *50*, 17–24. [CrossRef]

89. Cook, M.T.; Tzortzis, G.; Charalampopoulos, D.; Khutoryanskiy, V.V. Microencapsulation of a synbiotic into PLGA/alginate multiparticulate gels. *Int. J. Pharm.* **2014**, *466*, 400–408. [CrossRef]

90. Varankovich, N.; Martinez, M.F.; Nickerson, M.T.; Korber, D.R. Survival of probiotics in pea protein-alginate microcapsules with or without chitosan coating during storage and in a simulated gastrointestinal environment. *Food Sci. Biotechnol.* **2017**, *26*, 189–194. [CrossRef]

91. Molan, A.L.; Flanagan, J.; Wei, W.; Moughan, P.J. Selenium-containing green tea has higher antioxidant and prebiotic activities than regular green tea. *Food Chem.* **2009**, *114*, 829–835. [CrossRef]

92. Jantarathin, S.; Borompichaichartkul, C.; Sanguandeekul, R. Microencapsulation of probiotic and prebiotic in alginate-chitosan capsules and its effect on viability under heat process in shrimp feeding. *Mater. Today Proc.* **2017**, *4*, 6166–6172. [CrossRef]

93. Darjani, P.; Hosseini Nezhad, M.; Kadkhodaee, R.; Milani, E. Influence of prebiotic and coating materials on morphology and survival of a probiotic strain of Lactobacillus casei exposed to simulated gastrointestinal conditions. *LWT* **2016**, *73*, 162–167. [CrossRef]

94. Liserre, A.M.; Ré, M.I.; Franco, B.D.G.M. Microencapsulation of Bifidobacterium animalis subsp. lactis in Modified Alginate-chitosan Beads and Evaluation of Survival in Simulated Gastrointestinal Conditions. *Food Biotechnol.* **2007**, *21*, 1–16. [CrossRef]

95. Khosravi Zanjani, M.A.; Tarzi, B.G.; Sharifan, A.; Mohammadi, N. Microencapsulation of probiotics by calcium alginate-gelatinized starch with chitosan coating and evaluation of survival in simulated human gastro-intestinal condition. *Iran. J. Pharm. Res.* **2014**, *13*, 843–852.

96. Corona-Hernandez, R.I.; Álvarez-Parrilla, E.; Lizardi-Mendoza, J.; Islas-Rubio, A.R.; de la Rosa, L.A.; Wall-Medrano, A. Structural Stability and Viability of Microencapsulated Probiotic Bacteria: A Review. *Compr. Rev. Food Sci. Food Saf.* **2013**, *12*, 614–628. [CrossRef]

97. Borges, J.; Mano, J.F. Molecular Interactions Driving the Layer-by-Layer Assembly of Multilayers. *Chem. Rev.* **2014**, *114*, 8883–8942. [CrossRef]

98. Cook, M.T.; Tzortzis, G.; Charalampopoulos, D.; Khutoryanskiy, V.V. Production and evaluation of dry alginate-chitosan microcapsules as an enteric delivery vehicle for probiotic bacteria. *Biomacromolecules* **2011**, *12*, 2834–2840. [CrossRef]

99. Pinto, G.L.D.; Campana-Filho, S.P.; de Almeida, L.A. *Frontiers in Biomaterials: Chitosan Based Materials and its Applications*; Bentham Science Publishers: Sharjah, UAE, 2017; ISBN 978-1-68108-485-5.

100. Singh, P.; Medronho, B.; Alves, L.; da Silva, G.J.; Miguel, M.G.; Lindman, B. Development of carboxymethyl cellulose-chitosan hybrid micro- and macroparticles for encapsulation of probiotic bacteria. *Carbohydr. Polym.* **2017**, *175*, 87–95. [CrossRef]

101. Li, X.Y.; Chen, X.G.; Sun, Z.W.; Park, H.J.; Cha, D.-S. Preparation of alginate/chitosan/carboxymethyl chitosan complex microcapsules and application in Lactobacillus casei ATCC 393. *Carbohydr. Polym.* **2011**, *83*, 1479–1485. [CrossRef]

102. Cook, M.T.; Tzortzis, G.; Khutoryanskiy, V.V.; Charalampopoulos, D. Layer-by-layer coating of alginate matrices with chitosan–alginate for the improved survival and targeted delivery of probiotic bacteria after oral administration. *J. Mater. Chem. B* **2013**, *1*, 52–60. [CrossRef]

103. Kamalian, N.; Mirhosseini, H.; Mustafa, S.; Manap, M.Y.A. Effect of alginate and chitosan on viability and release behavior of Bifidobacterium pseudocatenulatum G4 in simulated gastrointestinal fluid. *Carbohydr. Polym.* **2014**, *111*, 700–706. [CrossRef]

104. Zhang, F.; Li, X.Y.; Park, H.J.; Zhao, M. Effect of microencapsulation methods on the survival of freeze-dried Bifidobacterium bifidum. *J. Microencapsul.* **2013**, *30*, 511–518. [CrossRef]

105. Fareez, I.M.; Lim, S.M.; Lim, F.T.; Mishra, R.K.; Ramasamy, K. Microencapsulation of Lactobacillus SP. Using Chitosan-Alginate-Xanthan Gum-β-Cyclodextrin and Characterization of its Cholesterol Reducing Potential and Resistance Against pH, Temperature and Storage. *J. Food Process Eng.* **2017**, *40*, e12458. [CrossRef]

106. Fareez, I.M.; Lim, S.M.; Mishra, R.K.; Ramasamy, K. Chitosan coated alginate–xanthan gum bead enhanced pH and thermotolerance of Lactobacillus plantarum LAB12. *Int. J. Biol. Macromol.* **2015**, *72*, 1419–1428. [CrossRef]

107. Yucel Falco, C.; Sotres, J.; Rascón, A.; Risbo, J.; Cárdenas, M. Design of a potentially prebiotic and responsive encapsulation material for probiotic bacteria based on chitosan and sulfated β-glucan. *J. Colloid Interface Sci.* **2017**, *487*, 97–106. [CrossRef]

108. Kanmani, P.; Satish Kumar, R.; Yuvaraj, N.; Paari, K.A.; Pattukumar, V.; Arul, V. Effect of cryopreservation and microencapsulation of lactic acid bacterium Enterococcus faecium MC13 for long-term storage. *Biochem. Eng. J.* **2011**, *58–59*, 140–147. [CrossRef]

109. Trabelsi, I.; Bejar, W.; Ayadi, D.; Chouayekh, H.; Kammoun, R.; Bejar, S.; Ben Salah, R. Encapsulation in alginate and alginate coated-chitosan improved the survival of newly probiotic in oxgall and gastric juice. *Int. J. Biol. Macromol.* **2013**, *61*, 36–42. [CrossRef]

110. Zaeim, D.; Sarabi-Jamab, M.; Ghorani, B.; Kadkhodaee, R.; Tromp, R.H. Electrospray assisted fabrication of hydrogel microcapsules by single- and double-stage procedures for encapsulation of probiotics. *Food Bioprod. Process.* **2017**, *102*, 250–259. [CrossRef]

111. Krasaekoopt, W.; Watcharapoka, S. Effect of addition of inulin and galactooligosaccharide on the survival of microencapsulated probiotics in alginate beads coated with chitosan in simulated digestive system, yogurt and fruit juice. *LWT Food Sci. Technol.* **2014**, *57*, 761–766. [CrossRef]

112. Prisco, A.D.; Maresca, D.; Ongeng, D.; Mauriello, G. Microencapsulation by vibrating technology of the probiotic strain Lactobacillus reuteri DSM 17938 to enhance its survival in foods and in gastrointestinal environment. *LWT Food Sci. Technol.* **2015**, *2*, 452–462. [CrossRef]

113. Ramos, V.; Albertengo, L.; Agullo, E. Present and Future Role of Chitin and Chitosan in Food. *Macromol. Biosci.* **2003**, *3*, 521–530.

114. Gandomi, H.; Abbaszadeh, S.; Misaghi, A.; Bokaie, S.; Noori, N. Effect of chitosan-alginate encapsulation with inulin on survival of Lactobacillus rhamnosus GG during apple juice storage and under simulated gastrointestinal conditions. *LWT Food Sci. Technol.* **2016**. [CrossRef]

115. Malmo, C.; La Storia, A.; Mauriello, G. Microencapsulation of Lactobacillus reuteri DSM 17938 Cells Coated in Alginate Beads with Chitosan by Spray Drying to Use as a Probiotic Cell in a Chocolate Soufflé. *Food Bioprocess Technol.* **2013**, *6*, 795–805. [CrossRef]

116. Kailasapathy, K. Survival of free and encapsulated probiotic bacteria and their effect on the sensory properties of yoghurt. *LWT Food Sci. Technol.* **2006**, *39*, 1221–1227. [CrossRef]

117. Champagne, C.P.; Fustier, P. Microencapsulation for the improved delivery of bioactive compounds into foods. *Curr. Opin. Biotechnol.* **2007**, *18*, 184–190. [CrossRef]

118. Le-Tien, C.; Millette, M.; Mateescu, M.-A.; Lacroix, M. Modified alginate and chitosan for lactic acid bacteria immobilization. *Biotechnol. Appl. Biochem.* **2004**, *39*, 347–354. [CrossRef]

119. Lee, J.S.; Cha, D.S.; Park, H.J. Survival of freeze-dried Lactobacillus bulgaricus KFRI 673 in chitosan-coated calcium alginate microparticles. *J. Agric. Food Chem.* **2004**, *52*, 7300–7305. [CrossRef]

120. Yu, W.-K.; Yim, T.-B.; Lee, K.-Y.; Heo, T.-R. Effect of skim milk-alginate beads on survival rate of bifidobacteria. *Biotechnol. Bioprocess Eng.* **2001**, *6*, 133–138. [CrossRef]

121. Urbanska, A.M.; Bhathena, J.; Prakash, S. Live encapsulated Lactobacillus acidophilus cells in yogurt for therapeutic oral delivery: preparation and in vitro analysis of alginate–chitosan microcapsules. *Can. J. Physiol. Pharmacol.* **2007**, *85*, 884–893.

122. Brinques, G.B.; Ayub, M.A.Z. Effect of microencapsulation on survival of Lactobacillus plantarum in simulated gastrointestinal conditions, refrigeration, and yogurt. *J. Food Eng.* **2011**, *103*, 123–128. [CrossRef]

123. Obradović, N.S.; Krunić, T.Ž.; Trifković, K.T.; Bulatović, M.L.; Rakin, M.P.; Rakin, M.B.; Bugarski, B.M. Influence of Chitosan Coating on Mechanical Stability of Biopolymer Carriers with Probiotic Starter Culture in Fermented Whey Beverages. Available online: https://www.hindawi.com/journals/ijps/2015/732858/abs/ (accessed on 28 February 2019).

124. García-Ceja, A.; Mani-López, E.; Palou, E.; López-Malo, A. Viability during refrigerated storage in selected food products and during simulated gastrointestinal conditions of individual and combined lactobacilli encapsulated in alginate or alginate-chitosan. *LWT Food Sci. Technol.* **2015**, *63*, 482–489. [CrossRef]

125. Chen, L.; Yang, T.; Song, Y.; Shu, G.; Chen, H. Effect of xanthan-chitosan-xanthan double layer encapsulation on survival of Bifidobacterium BB01 in simulated gastrointestinal conditions, bile salt solution and yogurt. *LWT Food Sci. Technol.* **2017**, *81*, 274–280. [CrossRef]

126. Nualkaekul, S.; Lenton, D.; Cook, M.T.; Khutoryanskiy, V.V.; Charalampopoulos, D. Chitosan coated alginate beads for the survival of microencapsulated Lactobacillus plantarum in pomegranate juice. *Carbohydr. Polym.* **2012**, *90*, 1281–1287. [CrossRef]

127. Talebzadeh, S.; Sharifan, A. Developing Probiotic Jelly Desserts with Lactobacillus Acidophilus. *J. Food Process. Preserv.* **2017**, *41*, e13026. [CrossRef]

128. Seyedain-Ardabili, M.; Sharifan, A.; Ghiassi Tarzi, B. The Production of Synbiotic Bread by Microencapsulation. *Food Technol. Biotechnol.* **2016**, *54*, 52–59. [CrossRef]

129. de Farias, T.G.S.; Ladislau, H.F.L.; Stamford, T.C.M.; Medeiros, J.A.C.; Soares, B.L.M.; Stamford Arnaud, T.M.; Stamford, T.L.M. Viabilities of Lactobacillus rhamnosus ASCC 290 and Lactobacillus casei ATCC 334 (in free form or encapsulated with calcium alginate-chitosan) in yellow mombin ice cream. *LWT* **2019**, *100*, 391–396. [CrossRef]

130. Zanjani, M.A.K.; Ehsani, M.R.; Tarzi, B.G.; Sharifan, A. Promoting Lactobacillus casei and Bifidobacterium adolescentis survival by microencapsulation with different starches and chitosan and poly L-lysine coatings in ice cream. *J. Food Process. Preserv.* **2018**, *42*. [CrossRef]

MDPI
St. Alban-Anlage 66
4052 Basel
Switzerland
Tel. +41 61 683 77 34
Fax +41 61 302 89 18
www.mdpi.com

MDPI Books Editorial Office
E-mail: books@mdpi.com
www.mdpi.com/books